W9-CTY-239

Coordination
Chemistry

DATE DUE

Dedication

This book is dedicated to Professor John Christian Bailar, Jr, our professional father and grandfather. Through his research and his training of students Professor Bailar is in large part responsible for the development of coordination chemistry in America.

Coordination Chemistry

Fred Basolo
Ronald C. Johnson

SCIENCE REVIEWS

British Library Cataloguing in Publication Data

Basolo, Fred
    Coordination chemistry.——2nd rev. and
    enlarged ed.
    1. Coordination compounds
    I. Title   II. Johnson, Ronald C.
    541.2'242   QD474
    ISBN 0-905927-47-8

Typeset by Mid-Country Press
Printed by Whitstable Litho Ltd

Contents

The authors

Fred Basolo has been Professor of Chemistry at Northwestern University since 1946 and is currently Morrison Professor of Chemistry. President of the American Chemical Society in 1982, member of the National Academy of Sciences, he is the author of over 300 scientific publications and has co-authored *Mechanisms of Inorganic Reactions* with R. G. Pearson. He has also edited and co-edited four other volumes in inorganic chemistry. The first edition of *Coordination Chemistry* has been translated into eight languages, including Japanese and Chinese.

Ronald C. Johnson is Professor of Chemistry at Emory University in Atlanta, Georgia where he has been on the faculty since 1961, he is a former President of the Georgia Academy of Science and has published both research papers and articles in the *Journal of Chemical Education*. Dr Johnson is author of *Introductory Descriptive Chemistry* and co-author with R. A. Day of *General Chemistry*.

Preface to the Second Edition

This second edition of *Coordination Chemistry* continues to be a basic presentation of this subject. It has been updated to cover a host of interesting new areas in which the ideas of coordination chemistry have impact. It includes a new last chapter which introduces organometallic chemistry and homogeneous catalysis and the rapidly expanding studies of metal ions in living systems and solids.

Coordination Chemistry is primarily concerned with metal complexes but many of its concepts are applicable to chemistry in general. Students just starting to study chemistry, therefore, will profit from an appreciation and understanding of the basic principles of coordination chemistry, which may be applied in more sophisticated fashion in advanced courses.

Although textbooks of general chemistry usually contain brief teatments of metal complexes and coordination chemistry, their limited space precludes the discussion of many important aspects of the subject. This being so, this book was written to more adequately present this subject to persons with a limited chemistry background. We believe it is appropriate for use by students who have had at least a year of high school chemistry. It would serve well as a supplement to introductory college chemistry courses, as part of the context of the second year of a high school chemistry course, as the subject of a special chemistry seminar, or as an introduction or review for graduate students, health professionals, or others who plan to do work which involves metal ions and complexes.

The authors are grateful for the many helpful comments from readers of the first edition of this book and would appreciate suggestions and reports of student reaction toward this edition. We wish to thank Dr S. A. Johnson who read the entire manuscript and made many helpful suggestions. One of us (F. B.) wishes to thank Dr V. Caglioti and the people in his Institute at the University of Rome, where part of the original writing of this book was done, for their generous help and hospitality.

Fred Basolo
Evanston, Illinois
March 1986

Ronald C. Johnson
Atlanta, Georgia

Periodic Table of the elements

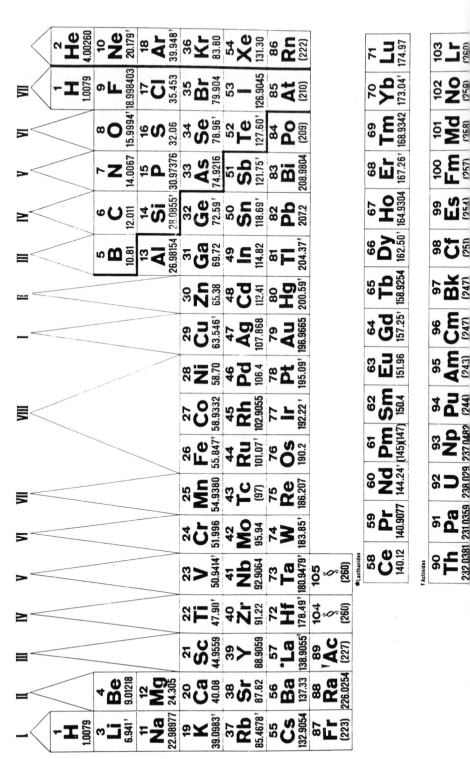

Chapter 1
Introduction and Historical Development

1.1 Introduction

At a very early stage in their course work in chemistry, students are introduced to a class of compounds referred to as *coordination compounds, metal complexes*, or just *complexes*. These compounds contain a central atom or ion, usually a metal, surrounded by several ions or molecules. The complex tends to retain its identity even in solution, although partial dissociation may occur. It may be a cation, an anion or be nonionic, depending on the sum of the charges of the central atom and the surrounding ions and molecules. It is the chemistry of this type of compound that is described in this book.

Coordination compounds play an essential role in chemical industry and in life itself. The importance of metal complexes becomes clear when one realizes that chlorophyll, which is vital to photosynthesis in plants, is a magnesium complex and that hemoglobin, which carries oxygen to animal cells, is an iron complex.

The nature and properties of metal complexes have been the subject of important research for many years and continue to intrigue some of the world's best chemists. One of the early Nobel prizes was awarded to Alfred Werner in 1913 for developing the basic concepts of coordination chemistry. The 1983 Nobel prize in chemistry was awarded to Henry Taube of Stanford University for his pioneering research on the mechanisms of inorganic oxidation–reduction reactions. He related rates of both substitution and redox reactions of metal complexes to the electronic structures of the metals, and made extensive experimental studies to test and support these relationships. His contributions are the basis for several sections in Chapter 6 and his concept of inner- and outer-sphere electron transfer is used by scientists worldwide.

Coordination compounds are used extensively in qualitative analysis as a means of separating certain metal ions and also as a means of positively identifying certain unknown ions. For example, you may have performed an experiment used to identify silver ion in solution. If silver ion is present, the addition of chloride ion gives an immediate white precipitate of silver chloride.

1

This precipitate dissolves in an excess of aqueous ammonia; but if an excess of nitric acid is added to this clear solution, the white precipitate forms again. This behavior is due to equilibria (1), (2).

$$Ag^+ + Cl^- \rightleftharpoons AgCl\downarrow \atop \text{white} \tag{1}$$

$$AgCl + 2NH_3 \rightleftharpoons [Ag(NH_3)_2]^+ + Cl^- \atop \text{clear solution} \tag{2}$$

A white precipitate forms, reaction (1), because AgCl is not soluble in water. It does, however, dissolve in excess NH_3 because the stable complex ion $[Ag(NH_3)_2]^+$ forms, reaction (2). The addition of excess HNO_3 to the clear solution causes equilibrium (2) to shift to the left, and the white precipitate of AgCl reappears. The reappearance is due to a lowering of the concentration of NH_3 owing to its reaction with H^+ to form NH_4^+.

The formation of metal complexes is often accompanied by very striking changes in color. One example is the use of aqueous solutions of $CoCl_2$ as invisible ink. This ink is essentially colorless; however, when the paper on which it was used is heated, the ink becomes a legible blue color. This color gradually fades as the paper absorbs atmospheric moisture. The phenomenon responsible for the appearance of color is shown by equilibrium (3).

$$[Co(H_2O)_6]Cl_2 \rightleftharpoons [CoCl_2(H_2O)_2] + 4H_2O \tag{3}$$
$$\text{pink} \qquad\qquad\qquad \text{blue}$$
$$\text{(colorless when dilute)}$$

The pink octahedral complex is almost colorless when dilute, so that writing done with it is practically invisible. Upon application of heat, water is driven off and a blue tetrahedral complex $[CoCl_2(H_2O)_2]$ is formed. Its color is sufficiently intense that the writing can easily be read. Upon standing, water is slowly taken up from the atmosphere and the original pink complex is regenerated, which makes the writing again invisible.

These examples illustrate that coordination compounds are common and are frequently encountered. Until the beginning of this century the nature of these materials was not understood, and the compounds were referred to as "complex compounds". This term is still used, but fortunately no longer in the same way. As a result of extensive research on such systems, our knowledge has increased so much that the systems are no longer considered complicated. In fact, a knowledge of the properties of complexes is necessary to obtain an understanding of the chemistry of metals.

1.2   Historical development

Scientific development usually comes about in a somewhat orderly fashion. The collection of facts by means of many carefully designed experiments is followed by an attempt to explain and correlate all facts with a suitable theory.

It must be remembered that, unlike facts, theories can and do often change as more information becomes available. The discussions that follow in this chapter and the next are good examples of how theories are modified or even at times completely discarded in order to accommodate new experimental results.

Discovery

It is difficult to state exactly when the first metal complex was discovered. Perhaps the earliest one on record is Prussian blue, $KCN \cdot Fe(CN)_2 \cdot Fe(CN)_3$, which was obtained by the artists' color maker Diesbach, in Berlin, at the beginning of the eighteenth century. However, the date usually cited is that of the discovery of hexaamminecobalt(III) chloride, $CoCl_3 \cdot 6NH_3$, by Tassaert (1798). This discovery marks the real beginning of coordination chemistry, because the existence of a compound with the unique properties of $CoCl_3 \cdot 6NH_3$ stimulated considerable interest in and research on similar systems. Although Tassaert's discovery was accidental, his realization that here was something new and different was certainly no accident, but a demonstration of his keen research ability. Serendipity accounts for many major breakthroughs in science; when a cure is found for cancer, for example, it may well come from an unexpected observation in a laboratory.

Tassaert's experimental observations could not be explained on the basis of the chemical theory available at that time. It was necessary to understand how $CoCl_3$ and NH_3, each a stable compound of presumably saturated valence, could combine to make yet another very stable compound. That they could so combine was a puzzle to chemists and a stimulus to further research, but the answer was not to be found until approximately 100 years later. During that time many such compounds were prepared and their properties studied. Several theories were proposed only to be discarded because they were inadequate to explain subsequent experimental data.

Preparation and properties

The preparation of metal complexes generally involves the reaction between a salt and some other molecule or ion (Chapter 4). Much of the early work was done with ammonia, and the resulting complexes were, and are, known as *metal ammines*. It was also found that anions such as CN^-, NO_2^-, NCS^-, and Cl^- form metal complexes. Many compounds were prepared from these anions, and at first each was named after the chemist who originally prepared it (Table 1.1). Some of these names are still used, but it soon become apparent that the system of nomenclature was not satisfactory.

Since many of the compounds are colored, the next scheme was to name compounds on the basis of color (Table 1.2). The reason behind this scheme was that colors of chloroammine complexes of cobalt(III) and chromium(III) containing the same number of ammonia molecules were found to be very

Table 1.1   Compounds named after their discoverers

Complex	Name	Present formulation
$Cr(SCN)_3 \cdot NH_4SCN \cdot 2NH_3$	Reinecke's salt	$NH_4[Cr(NCS)_4(NH_3)_2]$
$2PtCl_2 \cdot 2NH_3$	Magnus's green salt	$[Pt(NH_3)_4][PtCl_4]$
$Co(NO_2)_3 \cdot KNO_2 \cdot 2NH_3$	Erdmann's salt	$K[Co(NO_2)_4(NH_3)_2]$
$PtCl_2 \cdot KCl \cdot C_2H_4$	Zeise's salt	$K[PtCl_3(C_2H_4)]$

Table 1.2   Compounds named according to their color

Complex	Color	Name	Present formulation
$CoCl_3 \cdot 6NH_3$	Yellow	Luteocobaltic chloride	$[Co(NH_3)_6]Cl_3$
$CoCl_3 \cdot 5NH_3$	Purple	Purpureocoblatic chloride	$[CoCl(NH_3)_5]Cl_2$
$CoCl_3 \cdot 4NH_3$	Green	Praseocobaltic chloride	*trans*-$[CoCl_2(NH_3)_4]Cl$
$CoCl_3 \cdot 4NH_3$	Violet	Violeocobaltic chloride	*cis*-$[CoCl_2(NH_3)_4]Cl$
$CoCl_3 \cdot 5NH_3 \cdot H_2O$	Red	Roseocobaltic chloride	$[Co(NH_3)_5H_2O]Cl_3$
$IrCl_3 \cdot 6NH_3$[a]	White	Luteoiridium chloride	$[Ir(NH_3)_6]Cl_3$

[a] This compound was called *luteo* because it contains six ammonia molecules, not because of its color (see text).

nearly the same. Later, the scheme was used to designate the number of ammonias without regard to color. For example, $IrCl_3 \cdot 6NH_3$ is white and not yellow as implied by the prefix *luteo*. Clearly, such a system was not practical, and it had to be abandoned. The nomenclature system now used is described at the end of this chapter.

The chloroammine complexes of cobalt(III) [and those of chromium(III)] not only exhibit a spectrum of colors but also differ in the reactivity of their chlorides. For example, addition of a solution of silver nitrate to a freshly prepared solution of $CoCl_3 \cdot 6NH_3$ results in the immediate precipitation of all three chloride ions. The same experiment with $CoCl_3 \cdot 5NH_3$ causes instant precipitation of only two chloride ions; the third chloride precipitates slowly on prolonged standing. The results of such studies are summarized in Table 1.3. These observations suggest that in $CoCl_3 \cdot 6NH_3$ and in $IrCl_3 \cdot 6NH_3$ all chlorides are identical, but in $CoCl_3 \cdot 5NH_3$ and $CoCl_4 \cdot 4NH_3$ there are two different kinds of chlorides. One type is similar to that in sodium chloride and is readily precipitated as silver chloride, whereas the other type is held more firmly and does not precipitate.

Another kind of experiment provides useful information about the number of ions present in solutions of different complexes. The greater the number of ions in a solution, the greater is the electrical conductivity of the solution. Therefore, a comparison of the conductivities of solutions containing the same concentrations of coordination compounds permits an estimate of the number

Table 1.3 Number of chloride ions precipitated as AgCl

Complex	Number of Cl^- ions precipitated	Present formulation
$CoCl_3 \cdot 6NH_3$	3	$[Co(NH_3)_6]^{3+}, 3Cl^-$
$CoCl_3 \cdot 5NH_3$	2	$[CoCl(NH_3)_5]^{2+}, 2Cl^-$
$CoCl_3 \cdot 4NH_3$	1	$[CoCl_2(NH_3)_4]^+, Cl^-$
$IrCl_3 \cdot 3NH_3$	0	$[IrCl_3(NH_3)_3]$

Table 1.4 Molar conductivity of platinum(IV) complexes

Complex	Molar conductivity (S)	Number of ions indicated	Present formulation
$PtCl_4 \cdot 6NH_3$	523	5	$[Pt(NH_3)_6]^{4+}, 4Cl^-$
$PtCl_4 \cdot 5NH_3$	404	4	$[PtCl(NH_3)_5]^{3+}, 3Cl^-$
$PtCl_4 \cdot 4NH_3$	229	3	$[PtCl_2(NH_3)_4]^{2+}, 2Cl^-$
$PtCl_4 \cdot 3NH_3$	97	2	$[PtCl_3(NH_3)_3]^+, Cl^-$
$PtCl_4 \cdot 2NH_3$	0	0	$[PtCl_4(NH_3)_2]$
$PtCl_4 \cdot NH_3 \cdot KCl$	109	2	$K^+, [PtCl_5(NH_3)]^-$
$PtCl_4 \cdot 2KCl$	256	3	$2K^+, [PtCl_6]^{2-}$

of ions in each complex compound. This type of information was obtained for several series of complexes; some data are presented in Table 1.4. The results show that as the number of ammonia molecules in compounds decreases, the number of ions also falls to zero and then increases again.

One other important early observation was that two or more complexes sometimes exist having the same chemical composition yet different chemical and physical properties. These compounds are called *isomers*: examples are the green and violet forms of $CoCl_3 \cdot 4NH_3$. The colors of the two isomers are not always so dramatically different; but physical and chemical properties do differ. For example, the α and β forms of $PtCl_2 \cdot 2NH_3$ are both a cream color, but they differ in solubility and in chemical reactivity.

Several hypotheses and theories were proposed to account for all of these experimental facts. We shall discuss one that was used rather extensively and then proved to be wrong. We shall also discuss the coordination theory of Werner, which has withstood the test of time and provides a suitable explanation for the existence and behavior of metal complexes.

Blomstrand–Jörgensen chain theory

The development of a structural theory for organic compounds predated that for coordination compounds; thus, at the time people began to consider the structure of complexes, the concept of the tetravalency of carbon and the formation of carbon–carbon chains in organic compounds was already well recognized. This concept had a marked influence on chemists of that time. No doubt it influenced Blomstrand, professor of chemistry at the University of Lund, in Sweden, who in 1869 proposed the chain theory to explain the existence of metal complexes.

Because it was felt that elements had only one type of valence, Blomstrand and his student Jörgensen, who was later professor at the University of Copenhagen, suggested there could be only three bonds to cobalt(III) in its complexes. Therefore, a chain structure was used to account for the additional six ammonia molecules in $CoCl_3 \cdot 6NH_3$ (structure I). The three chlorides are separated by some distance from cobalt and were therefore believed to precipitate readily as silver chloride on addition of Ag^+. The theory

$$
\begin{array}{ll}
\underset{\diagup}{NH_3-Cl} & \underset{\diagup}{Cl} \\
Co-NH_3-NH_3-NH_3-NH_3-Cl & Co-NH_3-NH_3-NH_3-NH_3-Cl \\
\underset{\diagdown}{NH_3-Cl} & \underset{\diagdown}{NH_3-Cl} \\
\qquad\qquad I & \qquad\qquad\qquad II
\end{array}
$$

represented $CoCl_3 \cdot 5NH_3$ as structure II. In this structure, one chloride is attached directly to cobalt, and it was presumed that this is the one that does not ionize and does not precipitate instantly as silver chloride. Structure III for $CoCl_3 \cdot 4NH_3$ is also in accord with experiments that show that two chlorides are held more firmly than the third.

The next member of this series, $CoCl_3 \cdot 3NH_3$, was represented as structure IV. From this structure one would predict that the chlorides would behave as

$$
\begin{array}{ll}
\underset{\diagup}{Cl} & \underset{\diagup}{Cl} \\
Co-NH_3-NH_3-NH_3-NH_3-Cl & Co-NH_3-NH_3-NH_3-Cl \\
\underset{\diagdown}{Cl} & \underset{\diagdown}{Cl} \\
\qquad\qquad III & \qquad\qquad\qquad IV
\end{array}
$$

they do in $CoCl_3 \cdot 4NH_3$. Professor Jörgensen, a very able experimentalist, did not succeed in preparing the cobalt compound but made instead the analogous iridium complex $IrCl_3 \cdot 3NH_3$. A solution of this compound did not conduct a current, nor did it give a precipitate upon addition of silver nitrate. Thus, Jörgensen had succeeded in showing that his chain theory could not be correct.

Werner's coordination theory

Our present understanding of the nature of metal complexes is due to the ingenious insight of Alfred Werner, professor of chemistry in Zurich and winner of a Nobel prize in 1913. In 1893, at age 26, he proposed what is now commonly referred to as "Werner's coordination theory". His theory has been a guiding principle in inorganic chemistry and in the concept of valence. Three of its most important postulates are:

(1) Most elements exhibit two types of valence: (*a*) primary valence (represented by a solid line, ———), and (*b*) secondary valence (represented by a dashed line, – – – –). In modern terminology, (*a*) corresponds to *oxidation state* and (*b*) to *coordination number*.

(2) Every element tends to satisfy both its primary and secondary valence.

(3) The secondary valence is directed toward fixed positions in space. [Note that this is the basis for the stereochemistry of metal complexes (Chapter 3).]

Let us return to experimental facts described earlier and see how Werner's coordination theory explains these postulates. Again it is convenient to use chloroamminecobalt(III) complexes. According to the theory, the first member of the series, $CoCl_3 \cdot 6NH_3$, is designated as structure V and formulated as $[Co(NH_3)_6]Cl_3$. The primary valence or oxidation state of cobalt(III) is 3. The three chloride ions satisfy the primary valence of cobalt; these ions that neutralize the charge of the metal ion use the primary valence. The secondary valence or *coordination number* of cobalt(III) is 6. The coordination number is the number of atoms or molecules directly attached to the metal atom. The ammonia molecules use the secondary valence. They are said to be *coordinated* to the metal, and they are called *ligands*. Ligands (in this case ammonia) are directly attached to cobalt(III), and they are said to be in the *coordination sphere* of the metal. Here cobalt(III) is already surrounded by six ammonia molecules, so chloride ions can not be accommodated as ligands and hence are farther from the metal ion and are present as ionic Cl^-. Thus, a solution of the complex conducts current equivalent to four ions, and the chloride ions are readily precipitated by Ag^+ as silver chloride.

V   VI

Returning now to the theory, one finds that Werner represented $CoCl_3 \cdot 5NH_3$ as structure VI. He did this in accord with postulate (2), which states that both the primary and secondary valence tend to be satisfied. In $CoCl_3 \cdot 5NH_3$, there are only five ammonia molecules to satisfy the secondary valence. Therefore, one chloride ion must serve the dual function of satisfying both a primary and a secondary valence. Werner represented the bond between such a ligand and the central metal by a combined dashed–solid line $====$. Such a chloride is not readily precipitated from solution by Ag^+. The complex cation $[Co(III)(NH_3)_5Cl]^{2+}$ carries a charge of $2+$ because $Co^{3+} + Cl^- = +3 - 1 = +2$. The compound $CoCl_3 \cdot 5NH_3$ is now formulated $[CoCl(NH_3)_5]Cl_2$.

Extension of this theory to the next member of the series, $CoCl_3 \cdot 4NH_3$, requires structure VII. Two chloride ions satisfy both primary and secondary valence; hence they are firmly held in the coordination sphere. In solution, the compound, therefore, dissociates into two ions, Cl^- and $[CoCl_2(NH_3)_4]^+$.

A very significant point arises with the next member of the series, $CoCl_3 \cdot 3NH_3$. Werner's theory requires that it be represented as structure VIII and formulated as $[CoCl_3(NH_3)_3]$. The Werner theory predicted that the complex

VII   VIII

would not yield Cl^- ion in solution. In contrast, the chain theory predicted the compound would dissociate to give one chloride ion. The experimental results (Tables 1.3 and 1.4), when finally obtained, showed compounds of the type $[M(III)Cl_3(NH_3)_3]$ do not ionize in solution. This proved the chain theory was wrong and supported the coordination theory.

Postulate (3) of Werner's theory deals specifically with the stereochemistry of metal complexes. Coordination theory correctly explains many of the structural features of coordination compounds. One particularly important contribution of the theory was its use to determine the structure of six-coordinate complexes. Before the discovery of X-rays, the spatial configuration of molecules was determined by comparing the number of known isomers with the number theoretically possible for several possible structures. By this method it was possible to prove that certain structures were not correct and to obtain evidence in support of (but not confirmation of) a particular configuration.

Such a procedure was used successfully by Werner to demonstrate that six-coordinate complexes have an octahedral structure. The method starts with the assumption that a six-coordinate system has a structure in which the six ligands are situated at positions symmetrically equidistant from the central atom. If it is further assumed that three of the more probable structures are (1) planar, (2) a trigonal prism, and (3) octahedral (Table 1.5), then it is possible to compare the number of known isomers with the number theoretically predicted for each of these structures. Such a comparison shows that for the second and third compounds in Table 1.5 planar and trigonal prism structures predict there should be three isomers. Instead, complexes of these types were

Table 1.5 Known isomers compared with the number theoretically possible for three different structures

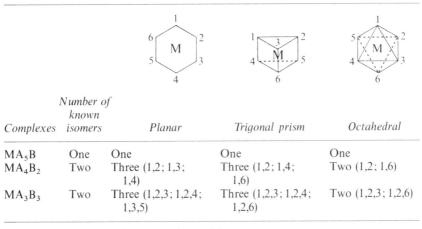

Complexes	Number of known isomers	Planar	Trigonal prism	Octahedral
MA$_5$B	One	One	One	One
MA$_4$B$_2$	Two	Three (1,2; 1,3; 1,4)	Three (1,2; 1,4; 1,6)	Two (1,2; 1,6)
MA$_3$B$_3$	Two	Three (1,2,3; 1,2,4; 1,3,5)	Three (1,2,3; 1,2,4; 1,2,6)	Two (1,2,3; 1,2,6)

Numbers in parentheses indicate positions of B groups.

found to exist only in two forms, in accord with the number of isomers theoretically possible for an octahedral structure.

The results furnish negative evidence, not positive proof, that planar and trigonal prism structures are incorrect. Failure to obtain a third isomer does not guarantee that complexes do not have these two structures; the third isomer may be much less stable or more difficult to isolate. Since negative evidence can only indicate what may be involved, the scientist is forced to design an experiment that gives positive evidence or proof. Werner was able to prove definitely that the planar and the trigonal prism structures cannot be correct. The proof involved demonstrating that complexes of the type $[M(AA)_3]$ are optically active, as discussed in Chapter 3.

1.3 Nomenclature

A comprehensive system of nomenclature of coordination compounds was not possible prior to Werner's coordination theory. As soon as it was realized that coordination compounds were either salts or nonionic species, it was possible to devise a systematic scheme for naming them. Salts were designed in the usual fashion, using a two-word name, and nonionic compounds were given a one-word name. For example, $[Co(NH_3)_6]Cl_3$ was called hexamminecobalti chloride and $[PtCl_2(NH_3)_2]$ was called dichlorodiammine-platino. According to this scheme the suffixes -a, -o, -i, and -e were used to designate the $+1$, $+2$, $+3$, and $+4$ oxidation states of the metal. This method has now been replaced by the Stock system of using Roman numerals in parentheses to indicate oxidation states. Thus $[Co(NH_3)_6]Cl_3$ is called hexaamminecobalt(III) chloride and $[PtCl_2(NH_3)_2]$ is diamminedichloro-platinum(II). The nomenclature system outlined below and that used throughout this text is recommended by the Inorganic Nomenclature Committee of the International Union of Pure and Applied Chemistry.

Order of listing ions

The cation is named first, and then the anion; this is the usual practice when naming a salt.

NaCl	sodium chloride
$[Cr(NH_3)_6](NO_3)_3$	hexaamminechromium(III) nitrate
$K_2[PtCl_6]$	potassium hexachloroplatinate(IV)

Nonionic complexes

Nonionic or molecular complexes are given a one-word name.

$[Co(NO_2)_3(NH_3)_3]$	triamminetrinitrocobalt(III)
$[Cu(CH_3COCHCOCH_3)_2]$	bis(acetylacetonato)copper(II)
$CH_3COCH_2COCH_3 = $ acetylacetone	

Names of ligands

Neutral ligands are named as the molecule; negative ligands end in o-; and positive ligands (of rare occurrence) end in -ium.

$NH_2CH_2CH_2NH_2$	ethylenediamine
$(C_6H_5)_3P$	triphenylphosphine
Cl^-	chloro
CH_3COO^-	acetato
$NH_2NH_3{}^+$	hydrazinium

Important exceptions to these results are water, ammonia, carbon monoxide, and nitric oxide.

H_2O	aquo
NH_3	ammine (Note the spelling with two m's; this applies only to NH_3; other amines are spelled with the usual one m.)
CO	carbonyl
NO	nitrosyl

Order of ligands

The ligands in a complex are listed in alphabetical order.

$[PtCl(NO_2)_2(NH_3)_3]_2SO_4$	triamminechlorodinitro-platinum(IV) sulfate

$NH_4[CrBr_4(en)]$ ammonium tetrabromo(ethylene-
                                         diamine)chromate(III)

en $= NH_2CH_2CH_2NH_2 =$ ethylenediamine

Numerical prefixes

The prefixes di-, tri-, tetra-, etc., are used before simple expressions such as bromo, nitro, and oxalato. Prefixes bis-, tris-, tetrakis-, etc., are used before complex names (chiefly expressions containing the prefixes mono-, di-, tri-, etc., in the ligand name itself) such as ethylenediamine and trialkylphosphine.

$K_3[Al(C_2O_4)_3]$ potassium trioxalatoaluminate(III)
$[CoCl_2(en)_2]_2SO_4$ dichlorobis(ethylenediamine)cobalt(III) sulfate

Termination of names

The ending for anionic complexes is -ate; alternatively, -ic if named as an acid. For cationic and neutral complexes the name of the metal is used without any characteristic ending.

$Ca_2[Fe(CN)_6]$ calcium hexacyanoferrate(II)
$[Fe(H_2O)_6]SO_4$ hexaaquoiron(II) sulfate
$[Ni(DMG)_2]$ bis(dimethylglyoximato) nickel(II)

$DMG = CH_3C$———$CCH_3 =$ dimethylglyoximate ion
                  ‖   ‖
                  N   N
          HO   O(−)

Oxidation states

The oxidation state of the central atom is designated by a Roman numeral in parentheses at the end of the name of the complex, without a space between the two. For a negative oxidation state a minus sign is used before the Roman numeral, and 0 is used for zero.

$Na[Co(CO)_4]$ sodium tetracarbonylcobaltate(−I)
$K_4[Ni(CN)_4]$ potassium tetracyanonickelate(0)

Bridging groups

Ligands that bridge two centers of coordination are preceded by the Greek letter μ, which is repeated before the name of each different kind of bridging group.

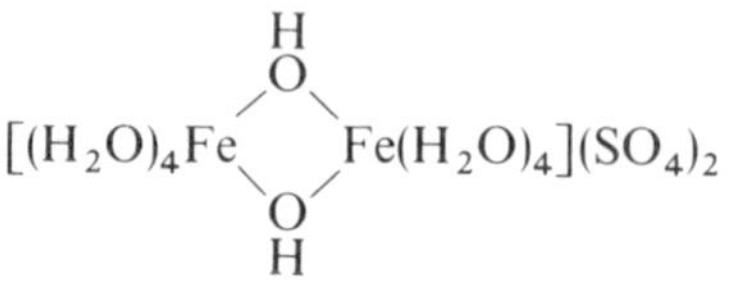

$$[(H_2O)_4Fe\underset{O\atop H}{\overset{H\atop O}{\diagup\hspace{-4pt}\diagdown}}Fe(H_2O)_4](SO_4)_2 \quad \mu\text{-dihydroxobis(tetraaquoiron(III)) sulfate}$$

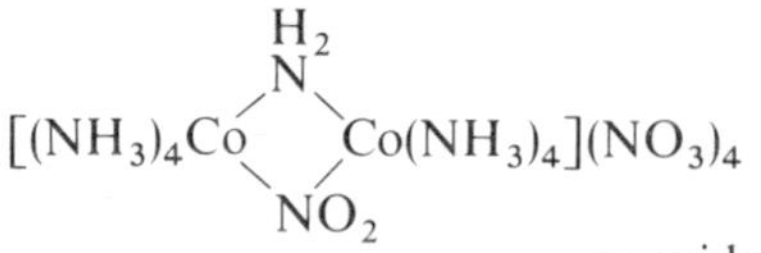

$$[(NH_3)_4Co\underset{NO_2}{\overset{H_2\atop N}{\diagup\hspace{-4pt}\diagdown}}Co(NH_3)_4](NO_3)_4$$

μ-amido-μ-nitrobis(tetraamminecobalt(III)) nitrate

Point of attachment

Whenever necessary, the point of attachment of a ligand is designated by placing the symbol (in italics) of the element attached after the name of the group, with separation by hyphen.

$(NH_4)_3[Cr(NCS)_6]$ ammonium hexathiocyanato-*N*-chromate(III)
$(NH_4)_2[Pt(SCN)_6]$ ammonium hexathiocyanato-*S*-platinate(IV)

For thiocyanate, nitrite, and cyanide ions the following optional names may be used:

$-SCN^-$, thiocyanato	$-NO_2^-$, nitro	$-CN^-$, cyano
$-NCS^-$, isothiocyanato	$-ONO^-$, nitrito	$-NC^-$, isocyano

Geometrical isomers

Geometrical and optical isomerism are discussed in Section 3.3. The naming of isomeric materials is included at this point for completeness. Geometrical isomers are generally named by the use of the terms *cis* to designate adjacent (90° apart) positions and *trans* for opposite (180° apart) positions. It is occasionally necessary to use a number system to designate the position of

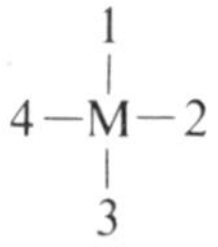

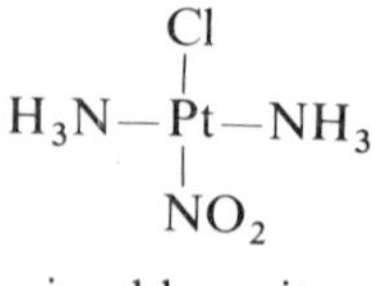

$$4-M-2 \atop {\overset{1}{\vert}\underset{3}{\vert}}$$

$$\underset{NO_2}{\overset{Cl}{H_3N-Pt-NH_3}}$$

trans-diamminechloronitroplatinum(II)

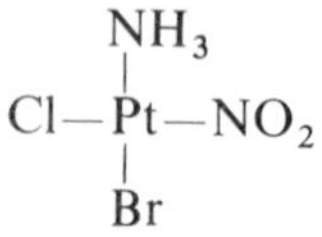

$$\underset{Br}{\overset{NH_3}{Cl-Pt-NO_2}}{}^-$$

1-ammine-3-bromochloronitroplatinate(II) ion

each ligand. For square planar complexes, groups 1–3 and 2–4 are in *trans* positions. Note that only two of the *trans* positions need be numbered in the name of the second complex. This is due to the fact that in a square complex the other two ligands must then be in *trans* positions. Since positions 2 and 4 are equivalent, these numbers need not be mentioned. Simple models are most helpful in visualizing these complexes.

The number system for octahedral complexes has the *trans* positions numbered 1–6, 2–4, and 3–5. An optional name for the second compound below is *trans*-ammineiodo-*trans*-bromochloronitro(pyridine)-platinum(IV).

$$
\begin{array}{c}
1 \quad 2 \\
| \quad .\!^{2} \\
5-M-3 \\
| \\
4 \quad 6
\end{array}
\qquad
\begin{array}{c}
NH_3\,.Br \\
| \\
H_3N-Rh-Br \\
| \\
H_3N \quad NH_3
\end{array}^{+}
$$

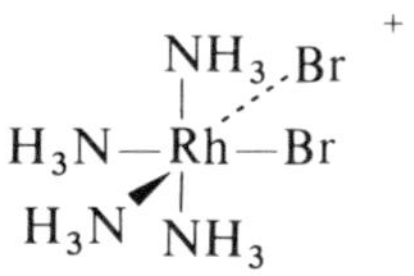

cis-tetraaminedibromorhodium(III) ion

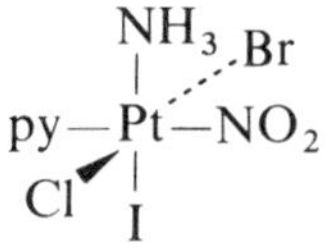

$$
\begin{array}{c}
NH_3\,.Br \\
| \\
py-Pt-NO_2 \\
| \\
Cl \quad I
\end{array}
$$

1-ammine-2-bromo-4-chloro-6-iodonitro-(pyridine)platinum(IV)

Optical isomers

If a solution rotates plane-polarized yellow light (the Na_D line) to the right, the solute is designated a (+) isomer; if to the left, a (−) isomer.

(+) $K_3[Ir(C_2O_4)_3]$ potassium (+)trioxalatoiridate(III)
(−) $[Cr(en)_3]Cl_3$ (−)tris(ethylenediamine)chromium(III) chloride

Abbreviations

Simple abbreviations are customarily used for complicated molecules in coordination compounds. Unfortunately, there is still no agreement on the abbreviation to use for a particular ligand. Some of the abbreviations that are commonly used and those which are used in this book are given in Table 1.6.

Miscellaneous terminology

It is convenient to include here the definition or description of some terms that have not yet been introduced. Ethylenediamine (en) occupies two coordination positions and thus behaves as if it were two molecules of ammonia tied together. Other molecules have the capacity to attach to the central atom at even more than two positions; for example, dien and EDTA may be attached

Table 1.6   Symbols used for some ligands

Symbol	Ligand name	Formula
en	ethylenediamine	$\ddot{N}H_2CH_2CH_2\ddot{N}H_2$
py	pyridine	

pn	propylenediamine	$\ddot{N}H_2CH_2CH(CH_3)\ddot{N}H_2$
dien	diethylenetriamine	$\ddot{N}H_2CH_2CH_2\ddot{N}HCH_2CH_2\ddot{N}H_2$
trien	triethylenetetramine	$\ddot{N}H_2CH_2CH_2\ddot{N}HCH_2CH_2\ddot{N}HCH_2CH_2\ddot{N}H_2$
bipy	2,2′-bipyridine	

| phen | 1,10-phenanthroline | |

| EDTA | ethylenediaminetetraacetato | |

| DMG | dimethylglyoximato | |

| gly | glycinato | $:NH_2CH_2COO\bar{\ }$ |
| acac | acetylacetonato | |

| PPh_3 | triphenylphosphine | $:P(C_6H_5)_3$ |
| Cp | cyclopentadienyl[a] | |

[a] Note this ligand is not given the usual o and ending of a negative ligand.

to three and six positions, respectively (Table 1.6). Such a group is called a *multidentate* or a *chelate* ligand. The correct usage of the adjective chelate is best illustrated by example. The salt $[Cu(en)_2]SO_4$, structure IX, is designated as a *chelate compound*, the cation as a *chelate ion*, and the ethylenediamine as a *chelate ligand*. The latter is also called a *bidentate* group or ligand. For greater numbers of points of attachment the terms used are: three, tridentate; four, quadridentate; five, quinquedentate; and six, sexidentate.

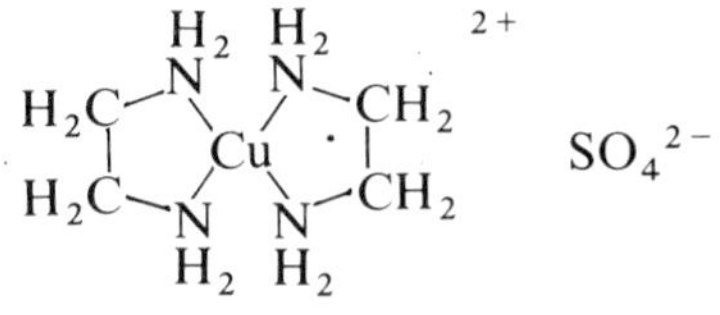

$$
\left[\begin{array}{c} \underset{\displaystyle H_2C}{\overset{\displaystyle H_2C}{}} \end{array} \right]
$$

IX

Whenever a ligand atom is attached to two metal ions, it is called a *bridging group* (see Section 1.3). The resulting complex is often called a "polynuclear complex", but a better term is *bridged complex*. This is preferred because the prefix "poly" usually denotes a high molecular weight, whereas these substances are often only dimers or trimers.

Problems

1. The compound $CoCl_3 \cdot 2en$ (en $= NH_2CH_2CH_2NH_2$) contains only one chloride ion that will be precipitated immediately upon addition of silver ion. (*a*) Draw the structure of this compound on the basis of the Blomstrand–Jörgensen chain theory of bonding. (*b*) Draw the structure on the basis of Werner's coordination theory. (*c*) Explain how each theory accounts for there being only one ionic chloride. (*d*) Explain why the chain theory cannot account for the stereochemistry of the compound.

2. Combinations of chromium(III), NH_3, SCN^-, and NH_4^+ can result in the formation of a series of seven coordination compounds, one of which is $[Cr(NH_3)_6](SCN)_3$. (*a*) Write the formulas for the other six members of the series. (*b*) Name each compound. (*c*) Indicate the complexes that should form geometric isomers (Section 3.3).

3. (*a*) Name each of the following compounds:

$[Co(NH_3)_6](NO_3)_3$
$[CrSO_4(NH_3)_5]ClO_4$

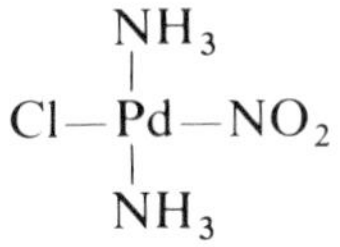

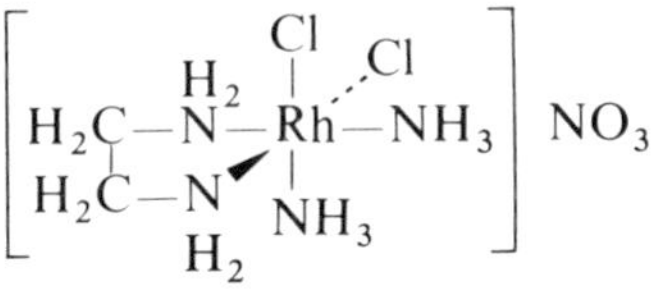

$[CrCN(H_2O)_5]SO_4$
$[CrNC(H_2O)_5]SO_4$
$Na_2[Fe(EDTA)]$
$K_4[Mo(CN)_8]$

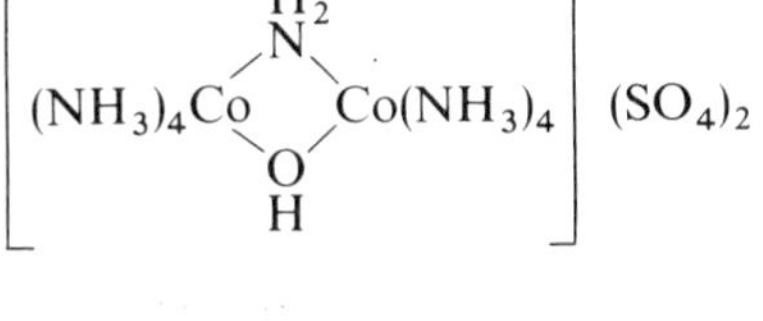

(*b*) Write the formula for each of the following compounds:

> tetraamminebromochlororuthenium(II)
> amminebis(ethylenediamine)iodoiridium(III) chloride
> strontium ethylenediaminedioxalatochromate(III)
> potassium tetrahydroxozincate(II)
> cesium hexafluoroiodate(V)
> μ-peroxobis(pentaamminecobalt(III)) sulfate
> *cis*-diglycinatonickel(II)
> sodium dithiosulfato-*S*-argentate(I)

4. Solid $CrCl_3 \cdot 6H_2O$ may be either $[Cr(H_2O)_6]Cl_3$, $[Cr(H_2O)_5Cl]Cl_2 \cdot H_2O$, or $[Cr(H_2O)_4Cl_2]Cl \cdot 2H_2O$. By making use of an ion-exchange column, it is possible to determine which of these three formulas is correct.

A solution containing 0.319 g of $CrCl_3 \cdot 6H_2O$ was passed through a cation-exchange resin in the acid form, and the acid liberated was titrated with a standard solution of NaOH. This required 28.5 ml of 0.125 M NaOH. Determine the correct formula of the chromium(III) complex.

References

Kauffman, G. B. (ed.) *Classics in Coordination Chemistry, Part I: The Selected Papers of Alfred Werner* (Dover, New York, 1968); Part 2 (1976); Part 3 (1978).
Nomenclature of Inorganic Chemistry—Definitive Rules 1970, 2nd edn (Butterworths, London, 1971).

Chapter 2

The Coordinate Bond

Werner's coordination theory, with its concept of secondary valence, provides an adequate explanation for the existence of such complexes as $[Co(NH_3)_6]Cl_3$. Some properties and the stereochemistry of these complexes are also explained by the theory, which remains the real foundation of coordination chemistry. Since Werner's work predated by about twenty years our present electronic concept of the atom, his theory does not describe in modern terms the nature of the secondary valence or, as it is now called, the *coordinate bond*. Three theories currently used to describe the nature of bonding in metal complexes are (1) valence bond theory (VBT), (2) crystal field theory (CFT), and (3) molecular orbital theory (MOT). We shall first describe the contributions of G. N. Lewis and N. V. Sidgwick to the theory of chemical bonding.

2.1 The electron–pair bond

In 1916, G. N. Lewis, professor of chemistry at the University of California in Berkeley, postulated that a bond between two atoms A and B can arise by their sharing a pair of electrons. Each atom usually contributes one electron. This electron pair bond is called a *covalent bond*. On this basis, he pictured the

$$\begin{array}{ccc} & H & & H \\ & \overset{..}{\underset{..}{H:C:H}} & \text{and} & :\overset{..}{\underset{..}{N}}:H \\ & H & & H \end{array}$$

molecules CH_4 and NH_3, respectively. This type of representation is now referred to as the *Lewis diagram* of a molecule.

An examination of the Lewis diagrams shows CH_4 and NH_3 to be similar in that there are two electrons (represented by dots) adjacent to each hydrogen, whereas carbon and nitrogen are each associated with eight electrons. An important difference is that one electron pair on nitrogen is not shared by a hydrogen. This permits ammonia to react in such a way as to share its "free" electron pair with some other atom. The resulting bond is also an electron pair or covalent bond; but because both electrons are furnished by the nitrogen, the bond is sometimes called a *coordinate covalent bond*.

The reaction of ammonia with acids to form ammonium salts, equation (1), produces a coordinate covalent bond. The four N—H bonds in NH_4^+ are, nonetheless, equivalent. This indicates that the distinction between coordinate and normal covalent bonds has little meaning. Ammonia may also share its

$$H^+ + \overset{\displaystyle H}{\underset{\displaystyle H}{:\!\ddot{N}\!:\!H}} \longrightarrow \left[\overset{\displaystyle H}{\underset{\displaystyle H}{H\!:\!\ddot{N}\!:\!H}} \right]^+ \tag{1}$$

free electron pair with substances other than hydrogen ion. When a metal ion takes the place of hydrogen ion, a metal ammine complex is formed, equations (2, (3), and (4). Since these reactions are generally carried out in aqueous

$$Ag^+ + \overset{\displaystyle H}{\underset{\displaystyle H}{:\!\ddot{N}\!:\!H}} \longrightarrow \left[\overset{\displaystyle H}{\underset{\displaystyle H}{Ag\!:\!\ddot{N}\!:\!H}} \right]^+ \overset{NH_3}{\longrightarrow} \left[\overset{\displaystyle H \qquad H}{\underset{\displaystyle H \qquad H}{H\!:\!\ddot{N}\!:\!Ag\!:\!\ddot{N}\!:\!H}} \right]^+ \tag{2}$$

$$Cu^{2+} + 4:NH_3 \longrightarrow \left[\overset{\displaystyle NH_3}{\underset{\displaystyle \ddot{N}H_3}{H_3N\!:\!\ddot{C}u\!:\!NH_3}} \right]^{2+} \tag{3}$$

$$Ni^{2+} + 6:NH_3 \longrightarrow \left[\begin{array}{c} \quad NH_3 \\ H_3\ddot{N}\!:\!\underset{\displaystyle Ni}{} \,:\!NH_3 \\ H_3\ddot{N}\!:\!\underset{\displaystyle \ddot{N}H_3}{} \,:\!NH_3 \end{array} \right]^{2+} \tag{4}$$

solution, it is more correct to indicate that the ions initially present are aquo complexes and that coordinated water is replaced by ammonia, equations (5) to (8).

$$[H\!:\!OH_2]^+ + :NH_3 \rightleftharpoons [H\!:\!NH_3]^+ + H_2O \tag{5}$$

$$[Ag(:OH_2)_2]^+ + 2:NH_3 \rightleftharpoons [Ag(:NH_3)_2]^+ + 2H_2O \tag{6}$$

$$[Cu(:OH_2)_4]^{2+} + 4:NH_3 \rightleftharpoons [Cu(:NH_3)_4]^{2+} + 4H_2O \tag{7}$$

$$[Ni(:OH_2)_6]^{2+} + 6:NH_3 \rightleftharpoons [Ni(:NH_3)_6]^{2+} + 6H_2O \tag{8}$$

These reactions are *Lewis acid–base reactions*. The Lewis theory of acids and bases defines an *acid* as a substance capable of accepting a pair of electrons and a *base* as a substance that donates a pair of electrons. The terms *acceptor* and *donor* are sometimes used for acid and base, respectively. A Lewis acid–base reaction results in the formation of a coordinate bond, equation (9).

$$\underset{\substack{\text{acid} \\ \text{(acceptor)}}}{A} \quad + \quad \underset{\substack{\text{base} \\ \text{(donor)}}}{:B} \quad \rightarrow \quad \underset{\substack{\text{coordinate} \\ \text{bond}}}{A\!:\!B} \tag{9}$$

The Lewis acid–base concept classifies metal ions as acids. Furthermore, compounds such as BF_3, $AlCl_3$, SO_3, and SiF_4, which can accept electron pairs, are also acids.

Ligands share electron pairs with metals; thus, ligands are Lewis bases. Some examples are the molecules $H_2O:$, $:NH_3$, $(C_2H_5)_3P:$, $:CO$, and $:NH_2CH_2CH_2\ddot{N}H_2$ and ions such as

$$:\ddot{\underset{..}{Cl}}:^- \quad :CN^- \quad :\ddot{\underset{..}{O}}H^- \quad :NO_3^- \quad :O-\overset{\overset{O}{\|}}{C}-\overset{\overset{O}{\|}}{C}-O:^{2-}$$

and

$$\begin{array}{cc} :OOCCH_2 & CH_2COO:^{4-} \\ \diagdown & \diagup \\ & \ddot{N}CH_2CH_2\ddot{N} \\ \diagup & \diagdown \\ :OOCCH_2 & CH_2COO: \end{array}$$

It is apparent why en and EDTA (Table 1.6) are able to function as bidentate and hexadentate ligands, respectively. Similarly, one can see that a ligand atom that contains more than one pair of free electrons may serve as a bridging atom, structure (I).

$$\begin{array}{ccc} (C_2H_5)_3P & \ddot{C}\ddot{l}\vdots & Cl \\ \diagdown & \diagup \diagdown & \diagup \\ & Pt \qquad Pt \\ \diagup & \diagdown \diagup & \diagdown \\ Cl & \vdots C\underset{..}{l}\vdots & P(C_2H_5)_3 \end{array}$$

I

2.2 Electronic structure of the atom

Before continuing the discussion of bonding theory, it is necessary to review briefly the electronic structure of the atom. Note that electrons in atoms are described as occupying orbitals which in turn constitute subshells and shells. Each orbital can hold no more than two electrons and electrons occupy the lowest energy orbitals of the atom. Figure 2.1 illustrates relative energies of atomic orbitals, where small circles represent orbitals. Note there is one s orbital in each shell, and three p orbitals, five d orbitals, and seven f orbitals in each shell.

Note that in the fourth shell the orbital energies have this order: $4s < 4p < 4d < 4f$. Within any shell the order of energies is $s < p < d < f$. However, also note that the $4s$ orbital is lower in energy than the $3d$ orbitals and that the $6s$ orbital is lower in energy than the $4f$ orbitals. Thus orbitals of one shell may be of higher energy than some of those in the next shell.

The energies of orbitals are influenced by the nuclear charge of the atom and by the nature and number of the other electrons in the atom. Therefore, in the potassium atom, the $3d$ orbital is of higher energy than the $4s$; in scandium the

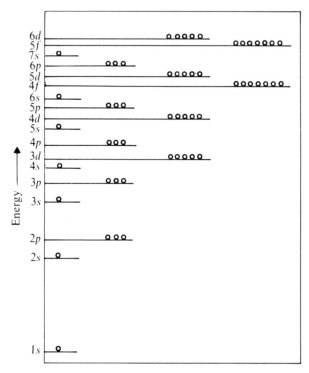

Figure 2.1 Energy-level diagram for an atom. Absolute energies of orbitals are distorted. Moreover, the ordering of orbitals is not the same for all atoms and ions.

$3d$ and $4s$ orbitals are of comparable energies; and in zinc the $4s$ orbital is of higher energy than the $3d$. The assignment of electronic configurations to atoms can only be approximated with a diagram such as Figure 2.1. The experimentally-observed configurations are presented in the Periodic Table at the end of this book.

Electrons fill each sublevel in accordance with *Hund's rule*, which states that electrons in orbitals of one sublevel tend to have the same spin. This means that electrons add to empty orbitals so long as they are available. This is reasonable, since electrons repel each other and would prefer to be in separate orbitals (as far from one another as possible). The electronic structure of N, Ti, and Mn can be designated as shown in Figure 2.2. Electrons in the p sublevel of N and in the d sublevel of Ti and Mn are unpaired. It is not necessary to list all of the sublevels as is done in the figure: only the electrons beyond those of the preceding noble gas (the *valence electrons*) are generally shown, since these are the ones involved in chemical bonding. One final point is that, for later use, it is more convenient to list the $3d$ sublevel before the $4s$, the $4d$ and $4f$ before the $5s$, etc.

Having briefly discussed the electronic structures of atoms, it is now

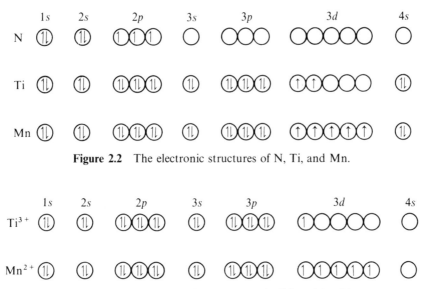

Figure 2.2 The electronic structures of N, Ti, and Mn.

Figure 2.3 The electronic structures of Ti^{3+} and Mn^{2+}.

necessary to consider the electronic structures of ions. In general, in the formation of positive ions electrons are lost from the highest-energy occupied orbitals of the atom. *In the case of the transition metals the outer s electrons are lost first.* Therefore, the electronic structures of Ti^{3+} and Mn^{2+} can be represented as shown in Figure 2.3. Note that Figure 2.1 would lead one to make the *incorrect* assumption that $3d$ electrons would be lost before $4s$ electrons.

Next it is necessary to be familiar with the shapes of these orbitals. By the "shape of an orbital" one means the shape of the region in space in which one is most likely to find an electron residing in that orbital. We shall limit ourselves to the s, p, and d orbitals, the ones most commonly involved in bond formation. The f orbitals may be used by the inner transition elements (the lanthanides and the actinides). The s orbital has spherical symmetry (Figure 2.4); the p orbitals have a dumbbell shape and each orbital is oriented along one of the three cartesian axes. The p_x orbital is oriented along the x axis, the p_y along the y axis, and the p_z along the z axis (Figure 2.5).

Four d orbitals have a cloverleaf shape, and one has a dumbbell shape with a sort of sausage wrapped around its center. Three of the cloverleaf orbitals, d_{xy}, d_{xz}, and d_{yz}, are oriented in the xy, xz, and yz planes, respectively, with their lobes situated between the two axes; the other cloverleaf orbital, $d_{x^2-y^2}$, is oriented in the xy plane and has its lobes along the x and y axes (Figure 2.6). The unique dumbbell-shaped orbital, d_{z^2}, is oriented along the z axis. To understand bonding in metal complexes, it is essential to maintain a mental picture of the three-dimensional shapes of these orbitals.

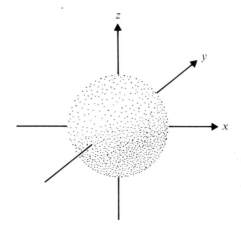

Figure 2.4   The spatial configuration of an *s* orbital.

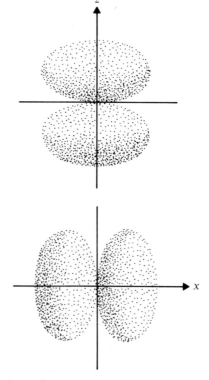

Figure 2.5   The spatial configurations of *p* orbitals.

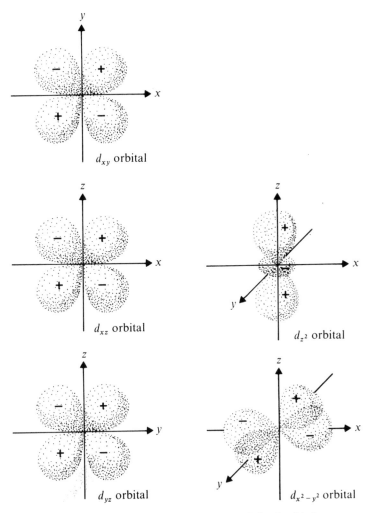

Figure 2.6 The spatial configurations of the d orbitals.

2.3 The concept of effective atomic number

The noble gas elements (He, Ne, Ar, Kr, Xe, and Rn) are unreactive; only recently have compounds of these elements been prepared. It has long been recognized that compounds in which each atom can, by electron sharing with other atoms, gather around itself a number of electrons equal to that found in a noble gas also tend to be very stable. Professor N. V. Sidgwick of Oxford University applied this observation to metal complexes. He postulated the central metal would surround itself with sufficient ligands so that the total number of electrons around the metal would be the same as that in the next noble gas. The number of electrons surrounding the coordinated metal is called its *effective atomic number* or EAN. For example, the EAN of cobalt(III) in $[Co(NH_3)_6]^{3+}$ is readily calculated as follows:

Co   atomic number 27, has 27 electrons
Co(III) $27 - 3 = 24$ electrons
$6(:NH_3)$[1] $2 \times 6 = 12$ electrons[1]
EAN of Co(III) in $[Co(NH_3)_6]^{3+} = 24 + 12 = 36$ electrons

Similarly determined EAN values for other metal complexes in many cases equal the atomic numbers of noble gases. There are, however, many exceptions to this rule; examples are $[Ag(NH_3)_2]^+$ and $[Ni(en)_3]^{2+}$, with EAN values of 50 and 38, respectively. This is unfortunate; for if the EAN of the central metal always exactly equaled the atomic number of a noble gas, then it would be possible to estimate the coordination number of metal ions.

One class of compounds that frequently does obey the EAN rule comprises the metal carbonyls and their derivatives. By using this rule it is possible to predict accurately coordination numbers of the simplest carbonyls and also predict whether the compounds can exist as monomers. For example, the EAN is 36 for the metals in the compounds $Ni(CO)_4$, $Fe(CO)_5$, $Fe(CO)_4Cl_2$, $Mn(CO)_5Br$, $CoNO(CO)_3$, and $Fe(NO)_2(CO)_2$. To estimate the EAN in these systems, it is convenient to say that, CO, Cl^-, and Br^- contribute two electrons and NO, three electrons. Manganese carbonyl has the formula $(CO)_5Mn-Mn(CO)_5$ which is the simplest formula possible if each Mn is to have an EAN of 36.

Electrons from each Mn $= 25$
Electrons from $5(:CO)$ $= 10$
Electron from $Mn-Mn$ bond$= \underline{1}$
                                                36

A Mn atom can gain one electron by forming a bond with another Mn atom. Each metal atom donates one electron to the bond, and each shares the two electrons.

Many chemists count only the electrons beyond the noble gas core to describe this same behavior. For example, in $Mn_2(CO)_{10}$ only the seven manganese electrons beyond the 18 in its argon noble gas core are counted.

Electrons from each Mn $= 7$
Electrons from $5(:CO)$ $= 10$
Electron from $Mn-Mn$ bond $= \underline{1}$
                          Total $= 18$

Each Mn in $Mn_2(CO)_{10}$ would therefore be called an eighteen-electron system. Eighteen-electron complexes have the EAN of a noble gas and the corresponding stability.

---

[1] Each ligand is considered to share two electrons with the central metal atom.

2.4   Valence bond theory

The valence bond theory was developed by Professor Linus Pauling, of the California Institute of Technology, and made available in his excellent book, *The Nature of the Chemical Bond,* published in 1940, 1948, and 1960. Along with the late Marie Curie, Professor Pauling is one of the very few persons to have been awarded two Nobel prizes, the Nobel prize in chemistry in 1954 and the Nobel peace prize in 1962. Pauling's ideas have had an important impact on all areas of chemistry; his valence bond theory has aided coordination chemists and has been extensively used. It can account reasonably well for the structure and magnetic properties of metal complexes. Extensions of the theory will account for other properties of coordination compounds such as absorption spectra, but other theories seem to do this more simply. Therefore, in recent years coordination chemists have favored the crystal field, ligand field, and molecular orbital theories.

It is useful to show the valence bond representations of the complexes $[CoF_6]^{3-}$ and $[Co(NH_3)_6]^{3+}$, which can then be compared with representations from the crystal field and molecular orbital theories to be discussed later. First, we must know from experiment that $[CoF_6]^{3-}$ contains four unpaired electrons, whereas $[Co(NH_3)_6]^{3+}$ has all of its electrons paired. Each of the ligands, as Lewis bases, contributes a pair of electrons to form a coordinate covalent bond. The valence bond theory designations of the electronic structures are shown in Figure 2.7. The bonding is described as being covalent. Appropriate combinations of metal atomic orbitals are blended together to give a new set of orbitals, called *hybrid orbitals.*

In six-coordinated systems, the hybrid orbitals involve the s, p_x, p_y, p_z, $d_{x^2-y^2}$, and d_{z^2} atomic orbitals. The resulting six sp^3d^2 or d^2sp^3 hybrid orbitals point toward corners of an octahedron. For $[CoF_6]^{3-}$ the d orbitals used have the same principal energy level as the s and p orbitals. A complex of the $nsnp^3nd^2$ type is called an *outer-orbital* complex because it uses "outer" d

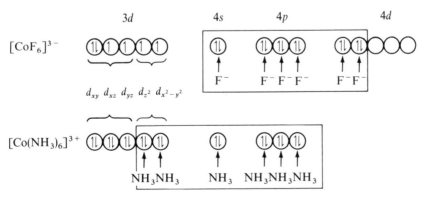

Figure 2.7   Valence bond theory representations of $[CoF_6]^{3-}$ and $[Co(NH_3)_6]^{3+}$.

orbitals. On the other hand, $[Co(NH_3)_6]^{3+}$ uses d orbitals of a lower principal energy level than the s and p orbitals. Such a complex, $(n-1)d^2nsnp^3$, is called an *inner-orbital* complex because it uses "inner" d orbitals. See Section 2.5 for nomenclature used in these systems on the basis of the crystal field theory (page 33).

2.5 Crystal field theory

Crystal field theory (CFT) has been widely used by chemists to explain the behavior of transition metal compounds. Although the more general ligand field and molecular orbital theories are more extensively used, the basic ideas of crystal field theory are still useful and provide a convenient bridge for understanding the more complicated theories. In direct contrast with valence bond theory (VBT), which is essentially a theory for covalent bonding, CFT is an ionic model. Metal ligand bonds are described as resulting from the attraction of positive metal ions for negatively charged ligands (or the negative end of uncharged but polar ligands). In CFT, covalent interactions are totally neglected. The greatest barrier to understanding CFT is frequently the inability of people to exclude covalent concepts such as electron pair bonds and overlapping orbitals from their thinking.

Calculations of coordinate bond energies can be made using classical potential energy equations that take into account the attractive and repulsive interactions between charged particles (10);

$$\text{bond energy} \propto \frac{q_1 q_2}{r} \tag{10}$$

q_1 and q_2 are charges on the interacting ions and r is the distance that separates ion centers. A similar equation applies to the interaction between an uncharged polar molecule and an ion. This approach gives results that are in reasonably good agreement with experimental bond energies for non-transition-metal complexes. For transition-metal complexes calculated values are often too small. This discrepancy is largely corrected when d orbital electrons are considered and allowance is made for the effect of ligands on the relative energies of d orbitals.

This refinement of electrostatic theory was first recognized and used by the physicists Bethe and Van Vleck in 1930 to explain colors and magnetic properties of crystalline solids. Their theory known as crystal field theory was proposed at about the same time as—or even a little earlier than—VBT, but it took about twenty years for the CFT to be recognized and used by chemists. Perhaps this was because CFT was written for physicists and VBT gave such a satisfying pictorial representation of the bonded atoms.

In 1951, several theoretical chemists working independently used CFT to interpret spectra of transition-metal complexes with such success that there followed an immediate avalanche of research activity in the area. It soon

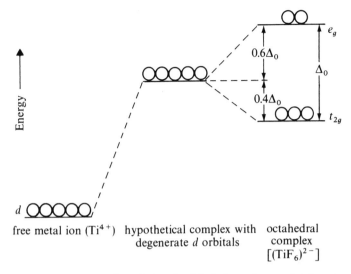

free metal ion (Ti^{4+}) hypothetical complex with octahedral
degenerate d orbitals complex
[(TiF$_6$)$^{2-}$]

Figure 2.8 The energies of the d orbitals in a free metal ion, in a hypothetical complex in which there is no crystal field splitting, and in an octahedral complex.

became apparent that CFT could explain in a semiquantitative fashion many of the properties of coordination compounds.

To understand CFT, it is necessary to have a clear mental picture of the spatial orientation of d orbitals (Figure 2.6). It is the interaction of the d orbitals of a transition metal with ligands surrounding the metal that produces crystal field effects. We can illustrate CFT by considering the octahedral complex $[\text{TiF}_6]^{2-}$. In a free Ti^{4+} ion, one isolated from all other species, the electronic configuration is $1s^2 2s^2 2p^6 3s^2 3p^6$; no d electrons are present. The five empty $3d$ orbitals of this ion have identical energies. This means that an electron may be placed in any one of these d orbitals with equal ease. Orbitals that have the same energy are called *degenerate orbitals*.

In $[\text{TiF}_6]^{2-}$, the Ti^{4+} ion is surrounded by six F$^-$ ions. These F$^-$ ions make it much more difficult to place electrons in the Ti^{4+} d orbitals due to repulsion of the electrons by the negative charge on F$^-$ ions. In other words, the energy of the d orbitals increases as F$^-$ ions (or other ligands) approach the orbitals (Figure 2.8). If the six F$^-$ ions surrounding Ti^{4+} in $[\text{TiF}_6]^{2-}$ were situated equally near each of the five d orbitals of Ti^{4+}, all of these d orbitals would have the same energy (they would be degenerate), but an energy considerably greater than that which they had in the free Ti^{4+} ion. However, an octahedral complex in which all d orbitals remain degenerate is but a hypothetical situation.

The complex $[\text{TiF}_6]^{2-}$ has an octahedral structure; for convenience, we shall visualize this complex with the six F$^-$ ions residing on the x, y, and z axes of a cartesian coordinate system (structure II). In this orientation, F$^-$ ions are very near the $d_{x^2-y^2}$ and d_{z^2} orbitals, which are referred to as e_g orbitals (Figure

$$z$$
$$\begin{array}{c} \text{F} \quad \text{F}^{\cdot\cdot} \\ \cdots\text{F}-\text{Ti}-\text{F}\cdots\rightarrow y \\ \text{F} \quad \text{F} \end{array}$$
$$x$$

II

2.8). These e_g orbitals point directly at the F^- ligands, whereas the d_{xy}, d_{xz}, and d_{yz} orbitals—called t_{2g} orbitals—point between the ligands.[1] Therefore, it is more difficult to place electrons in e_g orbitals than in t_{2g} orbitals, which means that e_g orbitals are of higher energy than t_{2g} orbitals. This conversion of the five degenerate *d* orbitals of the free metal ion into groups of *d* orbitals having different energies is the primary feature of CFT; it is known as *CF splitting*. As we have seen, splitting results because *d* orbitals have a certain orientation in space and because neighboring atoms, ions, or molecules can change the energy of orbitals that are directed toward them in space.

Many students find CFT and its concept of crystal field splitting difficult to visualize. The preceding discussion attempts to describe the essential concepts in simple terms on the basis of the spatial geometries of the *d* orbitals. This is the correct approach to CFT. It may, however, be helpful to develop a simple physical picture of crystal field splitting. Refer to Figure 2.9 and imagine that the metal ion with its electron cloud can be represented by a sponge ball. Now consider what happens when a rigid spherical shell (corresponding to ligands) is forced around the outside of the ball. The volume of the ball decreases, and the system has a higher energy, as is evident from the fact that the sponge will expand spontaneously to its original volume upon removal of the constricting shell. This change in energy corresponds to the increase in energy that results

---

[1] The symbols e_g and t_{2g} are terms used in the mathematical theory of groups. The *t* refers to a triply degenerate set of orbitals; the *e*, a doubly degenerate set.

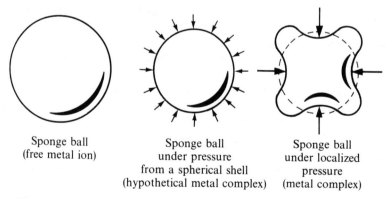

| Sponge ball (free metal ion) | Sponge ball under pressure from a spherical shell (hypothetical metal complex) | Sponge ball under localized pressure (metal complex) |

Figure 2.9 Crystal field effects visualized as a sponge ball under spherical pressure and under localized pressure. Compare with Figure 2.8.

from repulsion between electrons in a metal ion and electrons of ligands in the hypothetical complex (Figure 2.8).

Now if the rigid shell is allowed instead to concentrate its total force on six particular spots (for example, the corners of an octahedron), then the sponge is pressed inward at these positions but bulges outward between them. Compared with the spherically constricted system, the sponge at the six points of high pressure is at a higher energy and at the bulges between is at a lower energy. This corresponds to crystal field splitting with bulges related to t_{2g} orbitals and points of depression related to e_g orbitals.

In the preceding discussion, it was noted that the energy of the d orbitals of a metal ion increases when ligands approach the ion. This, in itself, suggests that a complex should be less stable than a free metal ion. However, the fact that complexes do form indicates that the complex is a lower energy configuration than the separated metal ion and ligands. The increase in energy of the d orbitals of the metal ion is more than compensated for by the bonding between metal ion and ligand. In the case of $[TiF_6]^{2+}$, bonding can be visualized as the result of electrostatic attraction between negative fluoride ions and the positive titanium ion.

In an octahedral arrangement of ligands, the t_{2g} and e_g sets of d orbitals have different energies. The energy separation between them is given the symbol Δ_o. It can be proved for an octahedral system that the energy of t_{2g} orbitals is $0.4\Delta_o$ less than that of the five hypothetical degenerate d orbitals that result if crystal field splitting is neglected (Figure 2.8). Therefore, the e_g orbitals are $0.6\Delta_o$ higher in energy than the hypothetical degenerate orbitals.

In an octahedral complex that contains one d electron (for example, $[Ti(H_2O)_6]^{3+}$) that electron will reside in the d orbital of lowest energy. Simple electrostatics does not recognize that d orbitals in a complex have different energies. Therefore, theory predicts that the d electron would have the energy of the hypothetical degenerate d orbitals. In fact, the d electron goes into a t_{2g} orbital that has an energy $0.4\Delta_o$ less than that of the hypothetical degenerate orbitals. Thus the complex will be $0.4\Delta_o$ more stable than the simple electrostatic model predicts. In simple terms, we can say that the d electron, and hence the whole complex, has a lower energy as a result of the placement of the electron in a t_{2g} d orbital that is as far from the ligands as possible. The $0.4\Delta_o$ is called the *crystal field stabilization energy* (CFSE) for the complex.

Table 2.1 gives CFSE's for metal ions in octahedral complexes. The values in Table 2.1 are readily calculated by assigning a value of $0.4\Delta_o$ for each electron put in a t_{2g} orbital and a value of $-0.6\Delta_o$ for each electron put in an e_g orbital. Thus, the CFSE for a d^5 system is either $3(0.4\Delta_o) + 2(-0.6\Delta_o) = 0.0\Delta_o$ or $5(0.4\Delta_o) + 0(-0.6\Delta_o) = 2.0\Delta_o$, depending on the distribution of the five electrons in the t_{2g} and e_g orbitals.

Simple electrostatics treats a metal ion as a spherical electron cloud surrounding an atomic nucleus. CFT provides a better model, since it admits that d electrons provide a nonspherical electron cloud in order to avoid positions in which ligands reside. (They provide a nonspherical electron cloud

Table 2.1   Crystal field stabilization energies for metal ions in octahedral complexes

d electrons in metal ions	t_{2g}	e_g	Stabilization, Δ_o	t_{2g}	e_g	Stabilization, Δ_o
1	↑ _ _	_ _	0.4			
2	↑ ↑ _	_ _	0.8			
3	↑ ↑ ↑	_ _	1.2			
4	↑ ↑ ↑	↑ _	0.6	↑↓ ↑ ↑	_ _	1.6
5	↑ ↑ ↑	↑ ↑	0.0	↑↓ ↑↓ ↑	_ _	2.0
6	↑↓ ↑ ↑	↑ ↑	0.4	↑↓ ↑↓ ↑↓	_ _	2.4
7	↑↓ ↑↓ ↑	↑ ↑	0.8	↑↓ ↑↓ ↑↓	↑ _	1.8
8	↑↓ ↑↓ ↑↓	↑ ↑	1.2			
9	↑↓ ↑↓ ↑↓	↑↓ ↑	0.6			
10	↑↓ ↑↓ ↑↓	↑↓ ↑↓	0.0			

by residing preferentially in low-energy orbitals that point between ligands.) Therefore, CFT explains why simple electrostatic calculations consistently underestimate the stability of transition-metal complexes and compounds; the simple approach neglects the nonspherical electron distribution and the resulting CFSE.

An early objection to a simple electrostatic treatment of bonding for metal complexes was that it could not explain the formation of square planar complexes. It was argued that if four negative charges are held to a positive central ion by electrostatic forces alone, then the negative charges must be at corners of a tetrahedron. Only in such a structure can negative groups attain maximum separation and so experience a minimum electrostatic repulsion. This is correct if the central ion is spherically symmetric. However, such symmetry is not typical of transition-metal ions, because electrons will reside in low-energy orbitals that point between ligands and do not have spherical symmetry. In Section 3.1 it is shown that CFT can account for square planar complexes and that it predicts the distortion of certain octahedral complexes.

We have considered the crystal field splitting for octahedral complexes; now let us consider complexes of other geometries. It is convenient to start with the crystal field splitting for an octahedral structure and consider how the splitting is affected by a change in geometry (Figure 2.10). In going from a regular octahedron to a square planar structure, the effect amounts to removal of any two *trans* ligands from the octahedron. Generally, we speak of the *xy* plane as the square plane, which means *trans* groups are removed from the *z* axis.

If ligands on the *z* axis are moved out so that the metal–ligand distance is only slightly greater than it is for the four ligands in the *xy* plane, the result is a tetragonal structure (Figure 2.10). This permits ligands in the *xy* plane to

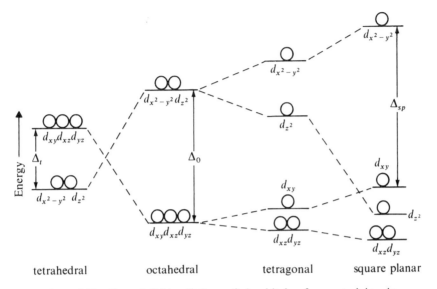

Figure 2.10 Crystal field splittings of d orbitals of a central ion in complexes having different geometries. The subscripts to Δ refer to the geometries.

approach the central ion more closely. Consequently, d orbitals in the xy plane experience a greater repulsion from ligands than they do in an octahedral structure, and we find an increase in energy of the $d_{x^2-y^2}$ and d_{xy} orbitals (Figure 2.10). At the same time d orbitals along the z axis or in the xz and yz plane experience a smaller repulsion from ligands, which are now some distance removed along the z axis. This results in a sizable decrease in energy for d_{z^2} orbital and a slight decrease for d_{xz} and d_{yz} orbitals, relative to the octahedral arrangement.

The same splitting pattern is found for a square pyramidal structure, in which there is one ligand on the z axis and the other four ligands plus the central atom are in the xy plane. The complete removal of two ligands on the z axis to give a square planar configuration is then accompanied by a further increase in energy of the $d_{x^2-y^2}$ and d_{xy} orbitals, as well as a further decrease for the d_{z^2}, d_{xz}, and d_{yz} orbitals.

The crystal field splitting of d orbitals for a tetrahedral structure is more difficult to visualize. We must first try to picture a tetrahedron placed inside a cube (Figure 2.11). The four corners of the tetrahedron are then located at four of the corners of the cube. If we now insert x, y, and z axes so they go through the center of the cube and protrude from the centers of its six faces, we can begin to see the position of the four ligands with respect to the d orbitals of the central atom. The d orbitals along the cartesian axes ($d_{x^2-y^2}$ and d_{z^2}) are further removed from the four ligands than are orbitals between the axes (d_{xy}, d_{xz}, and d_{yz}). Therefore, e_g orbitals ($d_{x^2-y^2}$ and d_{z^2}) are the low-energy d orbitals in

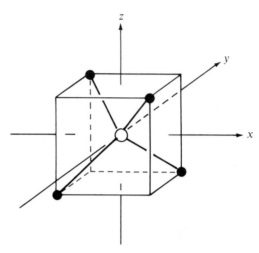

Figure 2.11 A tetrahedral complex with its center at the center of a cube.

tetrahedral complexes; t_{2g} orbitals (d_{xy}, d_{xz}, d_{yz}) are of relatively higher energy. It has been observed that the energy separation between e_g and t_{2g} orbitals, the crystal field splitting Δ_t, is only about one-half of Δ_o. Hence, crystal field effects favor the formation of octahedral complexes over that of tetrahedral complexes.

Magnetic properties

The magnetic properties of transition-metal complexes can readily be understood in terms of CFT. Transition metal ions have partially filled d orbitals. If these orbitals are degenerate, Hund's rule predicts that unpaired electrons will be present. For example, a metal ion containing three d electrons (called a d^3 system) should have three unpaired electrons (① ① ① ○ ○); a d^8 metal ion should have two unpaired electrons and three pairs of electrons (⑪ ⑪ ⑪ ① ①)). Materials that contain unpaired electrons are attraced to a magnet and are said to be *paramagnetic*. (This attraction is much weaker than that exhibited by ferromagnetic materials such as iron.) The magnitude of attraction of a material to a magnet is a measure of the number of unpaired electrons present.

Paramagnetism can be measured with a relatively simple device called a Gouy balance. The sample is placed in a tube suspended from a balance, and the weight of the sample is measured both in the presence and in the absence of a magnetic field. If the material is paramagnetic, it will weigh more while the magnetic field is present and attracting it. The increase in weight is a measure of the number of unpaired electrons in the compound.

A gaseous Co^{3+} ion, a d^6 system, has five degenerate d orbitals and is expected to have four unpaired electrons. However, some d^6 cobalt(III) complexes such as $[Co(NH_3)_6]^{3+}$ are not attracted to a magnet (they are *diamagnetic*). Complexes in which some of the unpaired electrons of the gaseous metal ion have been forced to pair are called *low-spin complexes*. The cobalt(III) complex $[CoF_6]^{3-}$ is paramagnetic and contains four unpaired electrons. It is an example of a *high-spin complex*. The electron distributions for these two complexes can be represented as ⑪ ⑪ ⑪ ◯ ◯ and ⑪ ① ① ① ①, respectively. A variety of names have been given to the behavior for which we have used the terms "high-spin" and "low-spin". These are summarized below.

$[Co(NH_3)_6]^{3+}$ ⑪ ⑪ ⑪ ◯ ◯ low-spin = spin-paired
 = inner-orbital complex

$[CoF_6]^{3-}$ ⑪ ① ① ① ① high-spin = spin-free
 = outer-orbital complex

It is now necessary to try to understand why in such systems *d* orbital electrons are distributed differently. It must be recognized that at least two effects determine the electron distribution. First, the normal tendency is for electrons to remain unpaired. Energy sufficient to overcome the repulsive interaction of two electrons occupying the same orbital is required to cause pairing. Second, in the presence of a crystal field, *d* orbital electrons will tend to occupy low-energy orbitals and thus avoid, as much as possible, repulsive interaction with ligands. If the stability thus gained (Δ) is large enough to overcome the loss in stability due to electron pairing, then electrons couple and the result is a low-spin complex. Whenever the crystal field splitting (Δ) is not sufficient, electrons remain unpaired and the complex is a high-spin type. Note in Figure 2.12 that the value of Δ_o for $[CoF_6]^{3-}$ is smaller than that for $[Co(NH_3)_6]^{3+}$. Complexes in which Δ is large will generally be low-spin complexes. Additional examples of crystal field splitting and electron distributions in metal complexes are shown in Figure 2.13.

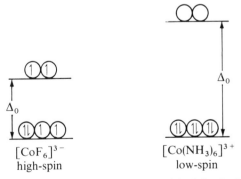

Figure 2.12 Relative crystal field splittings (Δ_o) of the *d* orbitals in high-spin and low-spin octahedral cobalt(III) complexes.

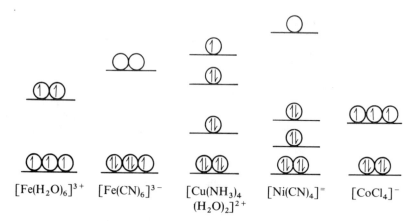

Figure 2.13 Crystal field splittings and electron distributions for some metal complexes. The structures of the first two complexes are octahedral, and the others (left to right) are tetragonal, square planar, and tetrahedral (see Figure 2.10).

The magnitude of the crystal field splitting determines whether d electrons in a metal ion will pair up. It also influences a variety of other properties of transition metals. The extent of the crystal field splitting depends on several factors. The nature of the groups (ligands) providing the crystal field is of greatest interest. From an electrostatic point of view, ligands with a large negative charge and those that can approach the metal closely (small ions) should provide the greatest crystal field splitting. Small, highly charged ions will make any d orbital they approach an energetically unfavorable place to put an electron. This reasoning is in agreement with the observation that the small F^- causes a greater crystal field splitting than the larger halide ions Cl^-, Br^-, and I^-.

Since crystal field splitting arises from a strong interaction of ligands with orbitals that point directly toward them and a weak interaction with those that point between, in order to achieve a large crystal field splitting it is desirable that a ligand "focus" its negative charge on an orbital. A ligand with one free electron pair (for example, NH_3) can be visualized as doing this much more readily than a species with two or more free electron pairs, see structures III and IV. This type of argument can be used to account for the observation

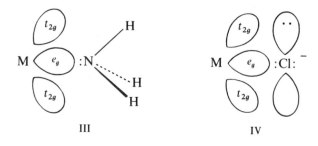

that neutral NH_3 molecules cause a greater crystal field splitting than H_2O molecules or negatively charged halide ions.

In general, however, it is difficult to explain the observed ability of various ligands to cause crystal field splitting with a simple electrostatic model. The crystal field splitting ability of ligands has been observed to decrease in the order (11). To account for this order, it is necessary to abandon a completely

strong-field ligands | intermediate-field ligands
$$CO, CN^- > phen > NO_2^- > en > NH_3 > NCS^- > H_2O > F^- >$$

weak-field ligands
$$RCO_2^- > OH^- > Cl^- > Br^- > I^- \qquad (11)$$

ionic electrostatic model for the bonding in complexes and to realize that covalent interactions also exist.

A modified CFT that includes the possibility of covalent bonding is called *ligand field theory*. It can account, at least qualitatively, for the crystal field splitting caused by various ligands. Molecules such as CO, CN^-, phen, and NO_2^-, which provide the largest crystal fields, are all able to form π *bonds* with the central metal atom (Section 2.6). This π bonding can markedly increase the magnitude of the crystal field splitting.

The crystal field splitting is also strongly influenced by the oxidation state of the metal ion and the type of d electrons present. In general, the higher the oxidation state of the metal ion, the larger will be the crystal field splitting. The complex $[Co(NH_3)_6]^{3+}$ is a diamagnetic low-spin complex, whereas $[Co(NH_3)_6]^{2+}$ is a paramagnetic high-spin complex. The crystal field splitting in the cobalt(III) complex is about twice as great as in the cobalt(II) complex; this results in pairing of electrons. One can attribute the larger Δ_o for cobalt(III) to the fact that ligands can approach more closely the slightly smaller, higher-charged metal ion and hence interact more strongly with its d orbitals. The crystal field splitting in $[Rh(NH_3)_6]^{3+}$ and $[Ir(NH_3)_6]^{3+}$ is greater than in $[Co(NH_3)_6]^{3+}$. In general, the crystal field splitting is greatest for complexes containing $5d$ electrons and least for those containing $3d$ electrons. One might attribute this behavior to the fact that $5d$ orbitals extend farther into space and thus interact more strongly with ligands.

Colors of transition metal ions

The greatest achievement of CFT is its success in interpreting colors of transition-metal compounds. One consequence of the comparatively small energy differences Δ between nonequivalent d orbitals in transition-metal complexes is that excitation of an electron from a lower to a higher level can be achieved by absorption of visible light. This causes the complex to appear colored. For example, an aqueous solution of titanium(III) is violet. The color is an indication of the absorption spectrum of the complex $[Ti(H_2O)_6]^{3+}$ (Figure 2.14). That the complex absorbs light in the visible region is explained by the electronic transition of the t_{2g} electron into an e_g orbital (Figure 2.15).

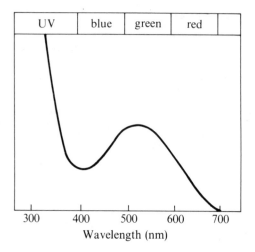

Figure 2.14 The absorption spectrum of $[Ti(H_2O)_6]^{3+}$. Solutions of $[Ti(H_2O)_6]^{3+}$ are red-violet because they absorb green light but transmit blue and red.

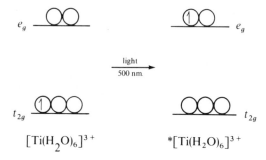

Figure 2.15 The *d–d* electronic transition responsible for the violet color of $[Ti(H_2O)_6]^{3+}$.

Absorption spectra of complexes containing more than one *d* electron are more complicated because a greater number of electronic transitons are possible.

Planck's equation (12) relates the energy *E* of an electronic transition to the wavelength λ of the light absorbed; *h* is Planck's constant (6.63×10^{-34} J-s),

$$E = \frac{hc}{\lambda} \qquad (12)$$

and *c* is the speed of light (3.00×10^8 m/s). The units of *E* are joules/molecule and of λ, meters.

From equation (12) it is possible to determine the energy difference Δ between *d* orbitals that are involved in the electronic transition. Collection of the constants h and c plus the use of Avogadro's number, 6.02×10^{23} molecules/mole, gives us equation (13). E has the units kilojoules/mole and λ

$$E = \frac{1.20 \times 10^5}{\lambda} \qquad (13)$$

is in nanometers. The maximum in the visible absorption spectrum of $[Ti(H_2O)_6]^{3+}$ is found at a wavelength of 500 nm, giving us a value of 240 kJ/mole for the energy difference between t_{2g} and e_g orbitals. The crystal field splitting Δ_o of 240 kJ is of the same order of magnitude as many bond energies. Although this value of 240 kJ is small compared to the heat of hydration of Ti^{3+} (14), 4300 kJ/mole, the crystal field splitting is very important and necessary to an understanding of transition-metal chemistry.

$$Ti^{3+} \text{ (gaseous)} + H_2O \rightarrow [Ti(H_2O)_6]^{3+} \text{ (aqueous)} + 4300 \text{ kJ/mole} \quad (14)$$

Note that the simple ionic model that is the basis of CFT does not accurately represent bonding in transition-metal compounds. There is ample experimental evidence that both ionic and covalent bonding play an important role. Nonetheless, the ionic CFT provides a simple model that will explain a great deal of transition-metal behavior and, moreover, one that has led and will lead to the formulation of many instructive experiments. The role of CFT in the structure, stability, and reactivity of complexes is discussed later.

2.6 Molecular orbital theory

The molecular orbital theory (MOT) is widely used by chemists. It includes both the covalent and ionic character of chemical bonds, although it does not specifically mention either. MOT treats the electron distribution in molecules in very much the same way that modern atomic theory treats the electron distribution in atoms. First, the positions of atomic nuclei are determined. Then orbitals around nuclei are defined; these molecular orbitals (MO's) locate the region in space in which an electron in a given orbital is most likely to be found. Rather than being localized around a single atom, these MO's extend over part or all of the molecule.

Since calculation of MO's from first principles is difficult, the usual approximate approach is the *linear combination of atomic orbitals* (LCAO) method. It seems reasonable that MO's of a molecule should resemble atomic orbitals (AO's) of the atoms of which the molecule is composed. From known shapes of AO's, one can approximate shapes of MO's. The linear combinations (additions and subtractions) of two atomic s orbitals to give two molecular orbitals are pictured in Figure 2.16. One MO results from addition of the parts of AO's that overlap, the other from their subtraction.

The MO that results from addition of two s orbitals includes the region in space between the two nuclei; it is called a *bonding MO*, and is of lower energy

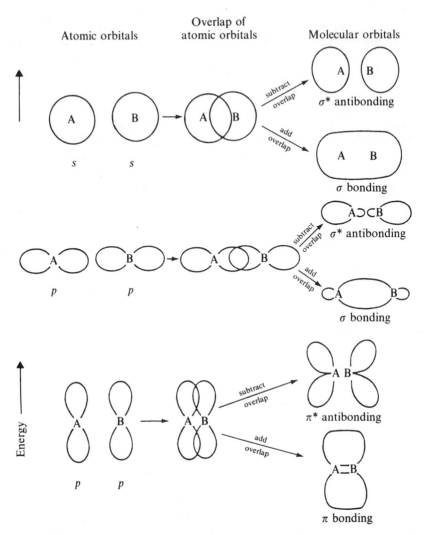

Figure 2.16   Formation of molecular orbitals by the LCAO method.

than either of the two *s* AO's from which it arose. The MO that results from subtraction of the parts of AO's that overlap does not include the region in space between the nuclei. It has a greater energy than the original AO's, and it is called an *antibonding MO*. The energy difference between bonding and antibonding MO's can be appreciated if one realizes that electrons, when they reside in a region between two nuclei, are strongly attracted by both nuclei.

Combinations of *s* atomic orbitals give *σ* (*sigma*) *MO's*. A combination of *p* AO's, as shown in Figure 2.16, may give either *σ* or *π* (*pi*) *MO's*. In a *π* MO, there is a plane passing through both nuclei along which the probability of

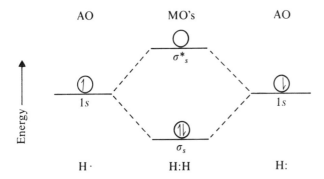

Figure 2.17 Molecular orbital diagram for the hydrogen molecule.

finding an electron is zero. Electrons in π MO's reside only above and below the bond axis.

To illustrate the use of MOT, let us look at MO energy diagrams for a few simple molecules. The H_2 molecule diagram is shown in Figure 2.17. In separated H atoms, one electron resides in each hydrogen AO. In the H_2 molecule, both electrons reside in the low-energy σ-bonding MO. The H_2 molecule is more stable than the separated H atoms; the two electrons are both in a lower-energy orbital in the molecule. The difference between the energy of the AO's and the bonding MO depends on how much the AO's overlap in the molecule. A large overlap results in a large difference and hence a *strong bond*; a small overlap results in a small difference, and hence the molecule will be of only slightly lower energy than the separated atoms.

The dihelium ion $He_2{}^+$ is a three-electron system; its MO energy level diagram is shown in Figure 2.18. Since an orbital can hold only two electrons, the third electron must go in the σ^* antibonding MO. This orbital is of higher energy than the AO's of separated He atoms; thus placing an electron in the σ^* MO represents a loss of energy and results in a less stable system. This is in agreement with the experimental observation that the $He_2{}^+$ bond energy is

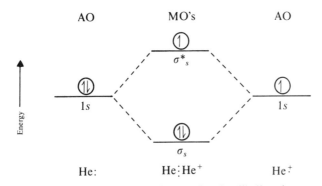

Figure 2.18 Molecular orbital diagram for the dihelium ion.

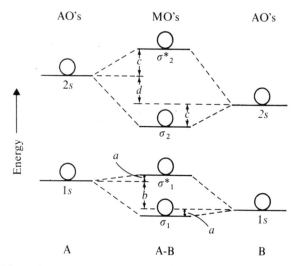

Figure 2.19 Molecular orbital diagram for an AB molecule.

only 238 kJ/mole compared with 436 kJ/mole for H_2. The four-electron He_2 molecule would be no more stable than two free He atoms.

A MO energy level diagram for the general molecule AB is shown in Figure 2.19. There are an infinite number of higher-energy MO's for the AB molecule just as these are an infinite number of higher-energy AO's for the A and B atoms, but the orbitals of interest are low-energy orbitals in which electrons reside. When two different types of atoms are present, the energies of the AO's are expected to differ (for example, $1s$ orbitals of A and B have different energies). The more electronegative element will have the lowest-energy AO's. The difference in energy between AO's of two elements (Figure 2.19b and d) is a measure of the amount of ionic character in the bond. In H_2, $1s$ orbitals of the two H atoms have the same energy; hence there is no ionic character in the bond.

The MO energy level diagrams for metal complexes are much more complicated than those for simple diatomic molecules. However, in the MO diagrams for $[Co(NH_3)_6]^{3+}$ and $[CoF_6]^{3+}$ in Figure 2.20 one can recognize several familiar features. On the left are $3d$, $4s$ and $4p$ atomic orbitals of Co^{3+}. The lower- and higher-energy AO's are of less interest. Since six ligands are involved, the right side of the diagram is somewhat different from diagrams we have seen previously. Only one orbital from each ligand, that used in σ bonding, is shown. (More complicated diagrams are sometimes used.) Since all six ligands are alike, the six ligand orbitals all have the same energy.

Ligand orbitals are, in general, of lower energy than the metal orbitals, and hence the bonds have some ionic character. Bonding MO's, in general, are closer in energy to ligand orbitals than to metal orbitals and are more like the ligand orbitals; placing metal electrons in these MO's thus transfers electronic charge from metal to ligands. Two d orbitals (the e_g orbitals, $d_{x^2-y^2}$ and d_{z^2}),

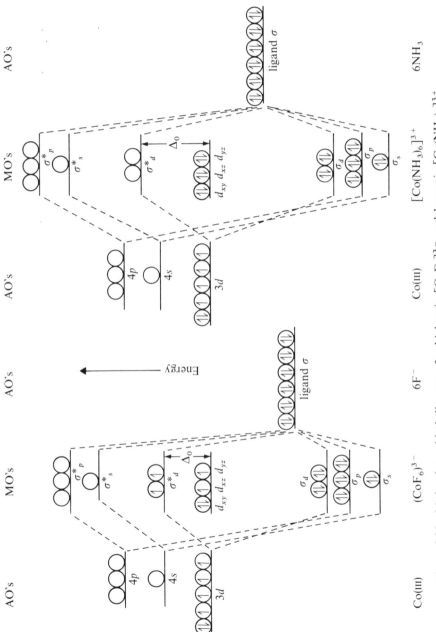

Figure 2.20 Molecular orbital diagrams for high-spin $[CoF_6]^{3-}$ and low-spin $[Co(NH_3)_6]^{3+}$.

the 4s, and the three 4p orbitals are oriented along the x, y, and z axes where the ligands are located. Therefore, orbital overlap with the ligand AO's results, and six bonding and six antibonding MO's are formed: σ_s (1), σ_p (3), σ_d (2), $\sigma_d{}^*$ (2), $\sigma_s{}^*$ (1), $\sigma_p{}^*$ (3). The t_{2g} (d_{xy}, d_{xz}, and d_{yz}) orbitals do not point at ligand orbitals and hence are not involved in σ bonding. Their energy is unchanged, and they are called nonbonding orbitals.

When cobalt(III) and ligand electrons are placed in the complex MO's, we find that the six bonding MO's are filled; this corresponds to six metal–ligand bonds. The remaining electrons are distributed among nonbonding MO's (the t_{2g} orbitals) and $\sigma_d{}^*$ (antibonding) MO's. The $\sigma_d{}^*$ MO's arise from the interaction of metal $d_{x^2-y^2}$ and d_{z^2} orbitals and ligand orbitals, but since the $\sigma_d{}^*$ MO's are nearer in energy to the metal $d_{x^2-y^2}$ and d_{z^2} orbitals, they do not differ markedly from them. Therefore, the placement of excess electrons in t_{2g} and $\sigma_d{}^*$ MO's is analogous to the arrangement predicted by the crystal field model, where the same number of electrons is distributed between t_{2g} and e_g orbitals.

If the difference in energy Δ_o between nonbonding t_{2g} orbitals and $\sigma_d{}^*$ MO is small, electrons remain unpaired; in $[\mathrm{CoF_6}]^{3-}$ this is what happens and d electrons are distributed $t_{2g}{}^4\sigma_d{}^{*2}$. The presence of two electrons in $\sigma_d{}^*$ orbitals effectively cancels the contribution of two electrons in bonding σ_d orbitals and hence weakens the Co—F bonds. When Δ_o is large as in $[\mathrm{Co(NH_3)_6}]^{3+}$, all electrons go into t_{2g} orbitals. The reasons for the energy separation between the t_{2g} and $\sigma_d{}^*$ or e_g orbitals are quite different in the two theories. According to CFT the crystal field splitting arises from electrostatic repulsion of d electrons by ligands. MOT essentially attributes the splitting to covalent bonding. The greater the overlap of e_g metal orbitals with ligand orbitals, the higher in energy will be the $\sigma_d{}^*$ orbital.

MOT can explain the influence of π bonds on the stability of metal complexes and on the magnitude of the crystal field splitting provided by ligands. Since a quantitative treatment of this subject is quite involved, only a qualitative explanation will be presented here. In the previous discussion, it was indicated that the strength of a covalent interaction depends on the extent of overlap of AO's on the two bonded atoms. In previous examples, only σ overlap was considered. In $[\mathrm{Fe(CN)_6}]^{4-}$ and a variety of other metal complexes, both σ and π bonding occur (Figure 2.21). In the σ bond, the ligand acts as a Lewis base and shares a pair of electrons with an empty e_g orbital. In the π bond, CN^- ion acts as a Lewis acid and accepts electrons from the filled t_{2g} orbital of the metal. The presence of π bonding as well as σ bonding strengthens the metal–ligand bond and contributes to the unusual stability of the $[\mathrm{Fe(CN)_6}]^{4-}$ ion. In oxyanions, such as $MnO_4{}^-$, σ and π bonding are also both important. In this case, the ligand (oxygen) provides the electrons for the π bond.

The large crystal fields that are provided by CN^-, CO, and other π-bonding ligands can be explained in this manner. The t_{2g} orbitals of a metal in an octahedral complex are oriented correctly for π bonding (Figure 2.21). As was

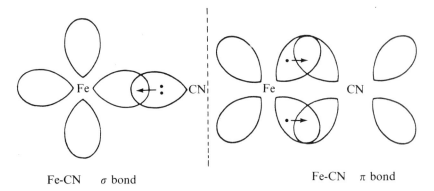

Fe-CN σ bond   Fe-CN π bond

Figure 2.21 σ and π bonds in $[Fe(CN)_6]^{4-}$. The π bond makes use of a filled d orbital of Fe^{2+} and an empty antibonding π^* orbital of CN^- (see π^* in Figure 2.16).

noted previously, the t_{2g} orbitals point between ligands and hence cannot form σ bonds. In a π bond with a ligand such as CN^-, t_{2g} electrons are partially transferred to the ligand. This process (a bonding interaction) lowers the energy of the t_{2g} orbitals. In Figure 2.8, one can see that a process that will lower the energy of t_{2g} orbitals must increase Δ_o.

The preceding discussion is a simplified MO approach to bonding, but it illustrates some of the basic ideas and a little of the usefulness of the theory. MOT is very effective in handling both covalent and ionic contributions to the metal ligand bond.

In conclusion, note that all three of these theories are, at best, only good approximations. All three can account qualitatively for many features of metal complexes; all three are used currently, and one or the other may be most convenient for a given application. The most versatile is MOT. Unfortunately, it is also the most complicated.

Problems

1. Determine the EAN of the metal in each of the following compounds. Note that several of these metals do not have an EAN equal to the atomic number of a noble gas.

$[PtCl_6]^{2-}$, $[Cr(H_2O)_6]^{3+}$, $[Ni(NH_3)_6]^{2+}$, $[Ni(CN)_4]^{2-}$, $[Ni(CN)_5]^{3-}$, $[AgCl_2]^-$, $[CoClNO_2(NH_3)_4]^+$, $Fe(CO)_5$, $MnNO(CO)_4$, $Co_2(CO)_8$, $[Al(OH)_4]^-$, $[Al(H_2O)_6]^{3+}$.

2. The crystal field splittings of d orbitals that arise from tetrahedral, tetragonal, and octahedral arrangements of ligands have been described.

Predict the splitting that will be produced for the following complexes and structures:

MX_2 linear complex

MX_6 four long $M-X$ bonds, two short MX bonds in an octahedral configuration

MX_6 planar structure with X groups at the corners of a hexagon

MX_8 a square prism structure (cube).

3. What is the CFSE for the following systems?

d^2 octahedral, d^5 high-spin octahedral, d^4 low-spin tetrahedral, d^2 tetrahedral, d^8 high-spin octahedral.

4. Represent the electronic configuration of the following complexes by using each of the three bonding theories, VBT, CFT, and MOT: $[Ni(H_2O)_6]^{2+}$ (high-spin), $[Fe(CN)_6]^{3-}$ (low-spin, neglect π bonding), $[Co(en)_3]^{3+}$ (low-spin), and $[PtCl_4]^{2-}$ (low-spin, VBT and CFT only).

References

Companion, A. L. *Chemical Bonding*, 2nd edn (McGraw-Hill, New York, 1979).

DeKock, R. and Gray, H. B. *Chemical Structures and Bonding* (Benjamin/Cummings, Reading, Massachusetts, 1980).

Gimarc, B. M. *Molecular Structure and Bonding* (Academic, New York, 1979).

Ballhausen, C. J. and Gray, H. B. *Molecular Electronic Structures, an Introduction* (Benjamin/Cummings, Reading, Massachusetts, 1980).

Pauling, L. *The Nature of the Chemical Bond*, 3rd edn (Cornell, Ithaca, New York, 1960).

Chapter 3

Stereochemistry

Stereochemistry is that branch of chemistry concerned with the structures of compounds. The liberal use of simple stick models is highly recommended as a visual aid to the study of three-dimensional structures. Stereochemistry is sometimes considered to be solely a province of organic chemistry, but this is a gross misapprehension. Because of the unique extent to which carbon forms carbon–carbon chains, organic compounds have a large variety of shapes and structures. However, if attention is focused on an individual carbon atom, the four groups surrounding it are located at the corners of a tetrahedron. Furthermore, since carbon is a second-row element, only the s and p orbitals are available for bond formation.

Inorganic stereochemistry deals with central atoms having coordination numbers from two to twelve. In inorganic compounds, it is often necessary to consider not only s and p orbitals but also d and even f orbitals. Isomerism, sometimes similar to that of organic compounds and sometimes different, is common in metal complexes.

3.1   Geometry of coordination compounds

Metal complexes have a variety of structures. Silver complexes are often linear; beryllium complexes are usually tetrahedral; iron forms a carbonyl compound that has a trigonal bipyramidal structure; cobalt(III) complexes are octahedral; and tantalum forms an eight-coordinated fluoride complex (Figure 3.1). Although a variety of coordination numbers and structures have been observed in metal complexes, the only common coordination numbers are four and six; the common structures corresponding to these coordination numbers are tetrahedral and square planar, and octahedral, respectively. In studying metal complexes, it soon becomes clear that the octahedral structure is by far the most common of these configurations.

An interesting and useful approach to the prediction of structure of molecules is given by the valence shell electron-pair repulsion (VSEPR). It, however, applies to molecules in which there are no valence d electrons, and is,

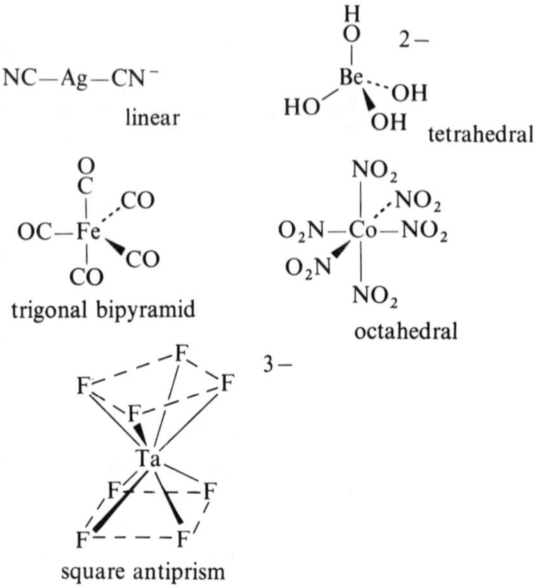

Figure 3.1 Illustrations of structures found in metal complexes.

therefore, not applicable to transition metal complexes. CFT provides the simplest explanation of the effect of d electrons on the structure of complexes.

CFT notes that d orbitals have a specific geometry and orientation in space and claims that d electrons will residue in the orbitals that are farthest from neighboring atoms or molecules. The presence of d electrons in six- or four-coordinate complexes may cause distortion of the expected octahedral or tetrahedral configuration. The distortion arises because ligands will avoid those areas around a metal ion in which the d electrons reside.

Let us consider the influence of metal d electrons on structure. If there are zero, five (unpaired), or ten d electrons present in the outer d subshell of an atom, there is no distortion of the structure of its complexes. This is true because empty, half-filled, and filled d subshells have spherical electrical symmetry; a charged particle (for example, a ligand) on a sphere having the metal at its center will encounter the same electrostatic force regardless of its position on the sphere. Therefore, the position that a ligand will occupy is not influenced by d electrons in these cases.

The complex $[\mathrm{Ti}(\mathrm{H_2O})_6]^{3+}$ contains one d electron; this electron will repel ligands that are near it. If this complex has the expected octahedral geometry, the one electron should be in a t_{2g} orbital, one which points between $\mathrm{H_2O}$ ligands. If the electron were in the d_{xy} orbital, one would expect the ligands in the xy plane to be repelled. This would lead to four long $\mathrm{Ti}-\mathrm{OH_2}$ bonds (and two short ones along the z axis). Four long bonds and two short ones would also be expected if the electron were placed in a d_{xz} or d_{yz} orbital. (The reader should convince himself of this.)

Since the t_{2g} orbitals point between ligands, one might expect that the effect of the presence of electrons in these orbitals would be small. In fact, unequal bond lengths in complexes such as $[Ti(H_2O)_6]^{3+}$ have not been detected by X-ray diffraction studies; however, the visible spectra of these complexes do provide evidence of a slight distortion. In octahedral d^3 complexes, such as $[Cr(H_2O)_6]^{3+}$, each t_{2g} orbital contains one electron. Each of the six ligands in an octahedral array would be near two of these d electrons, and hence all would experience the same repulsion. No distortion is expected or observed.

In $[Cr(H_2O)_6]^{2+}$, which is a d^4 high-spin system, the first three electrons go in t_{2g} orbitals and produce no distortion of an octahedral structure. The fourth electron goes in an e_g orbital that points directly at ligands. If the electron resides in a d_{z^2} orbital, the ligands on the z axis are repelled; if it resides in the $d_{x^2-y^2}$ orbital, the four ligands in the xy plane are repelled. In fact, six-coordinated d^4 metal complexes have distorted structures in all cases studied. For example, in MnF_3 each manganese(III) is surrounded by six F^- ions so arranged that four are closer to the Mn^{3+} ion than are the other two (Figure 3.2).

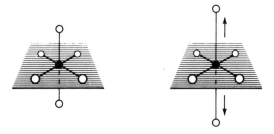

Figure 3.2 An example of a Jahn–Teller distortion. An octahedral molecule distorts to a tetragonal shape.

We have now considered the distortion of octahedral structures caused by the presence of 0, 1, 2, 3, 4, 5 (unpaired), and 10d electrons. It should be clear that high-spin d^6, d^7, d^8, and d^9 systems are similar to d^1, d^2, d^3, and d^4 systems, respectively. Six-coordinated complexes of d^9 metal ions exhibit distortions similar to those of d^4 complexes. The most common examples are copper(II) complexes. In $[Cu(NH_3)_4]^{2+}$, the tetragonal distortion is so marked that the square planar tetraammine complex results. Note, however, that solvent molecules occupy positions above and below the plane in solutions of complexes of this type; these solvent molecules are farther from the metal ion than are groups in the square plane. The distortion of symmetrical structures resulting from partially filled electronic energy levels (in this case the d sublevel) are called *Jahn–Teller* distortions.

The distortions of octahedral structure observed in important low-spin configurations should also be considered. Low-spin d^6 systems are similar to d^3 complexes. The six electrons completely fill the t_{2g} orbitals. Since each of the six ligands is close to two of these orbitals there is no tendency for distortion and regular octahedral structures are observed. Low-spin d^8 complexes are similar to d^4 systems. The last two electrons go into one e_g orbital and interact strongly with ligands that face this orbital. Marked distortions occur such that two ligands are much farther removed from the central metal than are the other four. In fact, low-spin d^8 complexes are almost invariably square planar. The distortions that result from the presence of d electrons in "octahedral" complexes are summarized in Table 3.1.

Table 3.1　The distortions of octahedral structures that result from the presence of d electrons

System	Predicted structure	Comments
High spin:		
d^1, d^6	Tetragonal distortion	Not observed
d^2, d^7	Tetragonal distortion	Not observed
d^3, d^8	No distortion	Experimentally verified
d^4, d^9	Large tetragonal distortion	Experimentally verified
d^5, d^{10}	No distortion	Experimentally verified
Low-spin:		
d^6	No distortion	Experimentally verified
d^8	Large tetragonal distortion	Square planar compounds

We have considered the distortions to octahedral structure that result from the presence of d electrons. Tetrahedral structures are also observed in metal complexes; however, they are less common than octahedral and distorted octahedral configurations. If four ligands surround a metal atom, a tetrahedral structure is expected. Two exceptions must be noted. As we have seen, four-coordinated low-spin d^8 complexes are square planar, as are many four-coordinated d^9 and high-spin d^4 complexes. Tetrahedral d^3, d^4, d^8, and d^9 systems should exhibit marked Jahn–Teller distortions; however, very few examples of this type of compound exist. Low-spin tetrahedral complexes need not be discussed, since there are no examples of such complexes. The tetrahedral crystal field splitting (Δ_t) is apparently too small to cause spin pairing.

Although it is possible to predict fairly accurately the stereochemistry of complex ions in which the coordination number of the central atom is known, it is much more difficult to predict the coordination number of the central atom. Large coordination numbers are favored by the electrostatic attraction of negatively charged ligands (or polar molecules) for a positive metal ion. Covalent bonding theories predict, in general, that the greater the number of bonds formed to an element, the greater is the stability of the resulting compound.

The tendency for large coordination numbers is opposed by steric and electrostatic (or Pauli) repulsion between ligands. No simple scheme has been presented which makes predictions from these criteria. Note, however, that the first-row transition elements are frequently six-coordinated. Four-coordination is observed primarily in complexes containing several large anions, such as Cl^-, Br^-, I^- and O^{2-}, or bulky neutral molecules.

The most common and important complex ions are hydrated metal ions. The coordination numbers and structures of some of these simple complexes have been determined. Isotope dilution techniques were used to show that Cr^{3+} and Al^{3+} are bonded rather firmly to six water molecules in aqueous solutions. The interpretation of the visible spectra of solutions of transition metal ions using CFT indicates that ions such as V^{2+}, Mn^{2+}, Fe^{2+}, Co^{2+}, Ni^{2+}, V^{3+}, Cr^{3+}, and Fe^{3+} are octahedral $[M(H_2O)_6]^{m+}$ species. For non-transition metal ions it has been more difficult to obtain structural information. However, nuclear magnetic resonance spectroscopy demonstrates that Be^{2+} in aqueous solution is surrounded by four water molecules. These data support the importance of six coordination. The only exception cited here is Be^{2+}, an element which obeys the octet rule.

3.2 Complexes with unusual structural features

In recent years, many examples of compounds with unusual coordination numbers, with novel structures, and with metal–metal bonds have been reported. Examples of complexes with odd coordination numbers and with coordination numbers greater than six are presented in Table 3.2.

The ion, structure I, occupies little space in the coordination sphere of a metal while acting as a bidentate chelate ligand and, therefore, can produce complexes in which the metal has a large coordination number. (The ligand is said to have a small "bite".) With the large thorium(IV) ion it yields the ten-

Table 3.2 Examples of compounds with unusual coordination numbers

Coordination number	Structure	Compound or ion
3	Planar	$[HgI_3]^-$
5	Trigonal bipyramid	$[Pt(SnCl_3)_5]^{3-}$
7	Pentagonal bipyramid	$[VO_2F_5]^{4-}$
7	Trigonal prism with one F^- at center of a square face	$[TaF_7]^{2-}$
8	Square antiprism	$[Zr(acac)_4]$
8	Dodecahedral	$[Mo(CN)_8]^{4-}$
9	Face-centered trigonal prism	$[ReH_9]^{2-}$
10	?	$[Th(O_2C_7H_5)_5]^{-b}$
12	Dodecahedral	$[Ce(NO_3)_6]^{3-a}$

[a] NO_3^- functions as a bidentate ligand.
[b] See structure I.

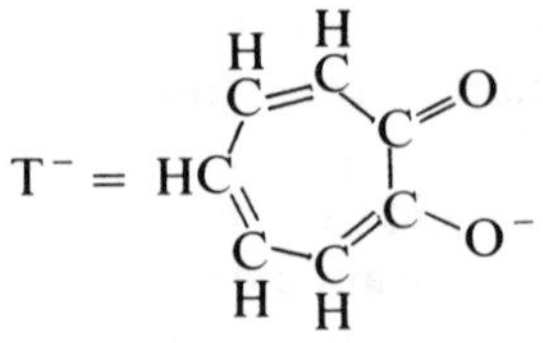

$$T^- = HC \begin{array}{c} H \ H \\ C=C \\ C=O \\ C \\ C-C \\ H \ H \end{array} C-O^-$$

I

coordinate complex $[Th(O_2C_7H_5)_5]^-$. With a variety of metal ions smaller than thorium(IV) eight-coordinate species are obtained.

Another striking achievement is the preparation of a six-coordinate complex with a non-octahedral structure. The trigonal prismatic geometry, structure II, was observed for $[Re(S_2C_2(C_6H_5)_2)_3]$. In this structure, the coordinated sulfurs are nearer to each other than in an octahedral arrangement and it has been suggested that there is bonding between adjacent sulfur atoms. The octahedral geometry found in most six-coordinate complexes results because ligands normally repel one another.

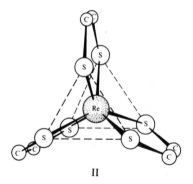

II

A variety of compounds with metal–metal bonds have been prepared. Mercurous compounds which contain the $[Hg–Hg]^{2+}$ unit have long been known. More recent examples include compounds in which carbon monoxide is a ligand such as $(Co)_5Mn–Re(CO)_5$ and $(C_6H_5)_3PAu–Co(CO)_4$. A number of complexes with transition metal tin bonds have been prepared. One of these, $[Pt(SnCl_3)_5]^{3-}$, contains five-coordinated platinum(II) and is an excellent hydrogenation catalyst.

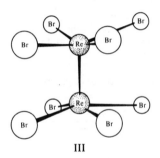

III

Another example $Br_4Re-ReBr_4$, structure III, has two unusual structural features. The bromines attached to the two different metal atoms are not staggered to minimize repulsive interactions, but are eclipsed. Moreover, the rhenium–rhenium bond is very short. These structural features and the magnetic properties have been interpreted as indicating that a quadruple bond links the metal atoms. The bonds include one σ-bond, two π-bonds, and a δ (delta)-bond. A δ bond can be visualized by placing the bonded atoms on a z coordinate axis and having overlap between two d_{xy} orbitals (one from each metal atom), structure IV.

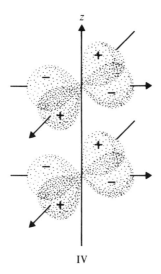

IV

It is now clear that a variety of elements form "clusters" of atoms which are held together, at least in part, by bonds between like atoms. Transition metals do this and the structures of some representative compounds are illustrated in Figure 3.3. Note that in $Ta_6Cl_{12}^{2+}$ the oxidation state of tantalum is $+7/3$. Non-integral oxidation states are also observed in other transition metal clusters. Some metal carbonyls also contain clusters of metal atoms; for example, in $Co_4(CO)_{12}$ there is a tetrahedron of cobalt atoms and in $Rh_6(CO)_{16}$ the six rhodium atoms are arrayed at the corners of an octahedron. A measure of the stability of some of these clusters is the fact that Co_4 units survive conditions in a mass spectrometer which strip all the CO's from a $Co_4(CO)_{12}$ molecule. At present, the metal cluster which contains the largest number of metal atoms, the champion, is $[H_2Pt_{38}(CO)_{44}]^{2-}$.

Since metal surfaces can be effective catalysts, much effort has been exerted to make metal clusters such as $[AuAu_8(PPh_3)_8]^{3+}$ (see Figure 3.3) which may mimic the behavior of a metal surface. It is much easier to study small clusters of metal atoms than a metal surface, especially if they can be dissolved in some liquid. Such studies should provide insight into how metal surfaces act as catalysts.

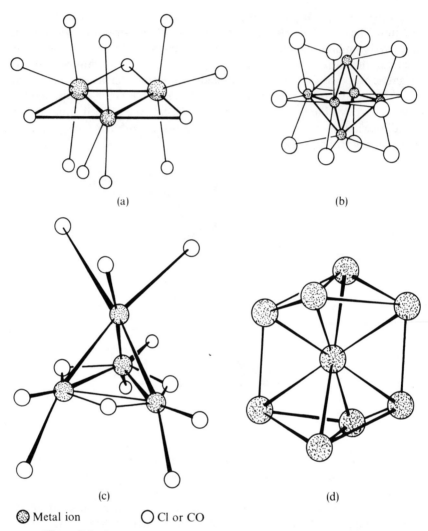

● Metal ion   ○ Cl or CO

Figure 3.3   Structures of metal clusters. (a) $Re_3Cl_{12}^{3-}$; (b) $Rh_6(CO)_{12}$; (c) $Co_4(CO)_{12}$; (d) $AuAu_8$ in $[AuAu_8(PPH_3)_8]^{3+}$.

The bonding in metal clusters is interesting and is best described using molecular orbitals which are delocalized over the cluster. The chemistry of boron contains numerous examples of clusters of boron atoms, but these are not normally considered in the realm of coordination chemistry.

3.3   Isomerism in metal complexes

Molecules or ions having the same chemical composition but different structures are called *isomers*. The difference in structure is usually maintained in solution. Metal complexes exhibit several different types of isomerism; the two most important are geometrical and optical. Other types will also be described, and specific examples will be given. Note that, in general, only complexes which react slowly are found to exhibit isomerism. Complexes that react rapidly often rearrange to yield only the most stable isomer (Chapter 6).

Geometrical isomerism

In metal complexes, ligands may occupy different positions around the central atom. Since the ligands in question are usually either next to one another (*cis*) or opposite each other (*trans*), this type of isomerism is often also referred to as *cis–trans isomerism*. Such isomerism is not possible for complexes with coordination numbers of 2 or 3 or for tetrahedral complexes. In those systems, all coordination positions are adjacent to one another. However, *cis–trans* isomerism is very common for square planar and octahedral complexes, the only two types to be discussed here. Methods of preparation and reactions of some of these compounds are described in Chapter 4.

Platinum(II) complexes are very stable and slow to react; among them are numerous examples of square planar geometrical isomers. The best known of these are *cis-* and *trans-*[$PtCl_2(NH_3)_2$], structures V and VI. In recent years,

$$
\begin{array}{cc}
\text{Cl} & \text{NH}_3 \\
| & | \\
\text{Cl}-\text{Pt}-\text{NH}_3 & \text{Cl}-\text{Pt}-\text{Cl} \\
| & | \\
\text{NH}_3 & \text{NH}_3 \\
\textit{cis} & \textit{trans} \\
\text{V} & \text{VI}
\end{array}
$$

the *cis* isomer has been extensively studied, since this molecule has had notable success as an anticancer drug. It is now marketed as *cis*-Platine, being an almost certain cure for testicular cancer.

The chemistry of platinum(II) complexes has been examined in detail, particularly by Russian chemists. Many compounds of the types *cis-* and *trans-*[PtX_2A_2], [PtX_2AB], and [$PtXYA_2$] are known. (A and B are neutral ligands such as NH_3, py, $P(CH_3)_3$, and $S(CH_3)_2$; X and Y are anionic ligands such as Cl^-, Br^-, I^-, NO_2^-, and SCN^-.) Isomers can be distinguished by X-ray diffraction and a variety of other techniques. Methods for determining the structures of geometrical isomers are discussed in Section 4.9.

A few compounds of platinum(II) containing four different ligands, [PtABCD], are known. Realizing that either B, C, or D groups may be *trans* to A, it is apparent that there are three isomeric forms for such a compound. The first complex of this type to be obtained in three forms was the cation [$PtNO_2NH_3(NH_2OH)py$]$^+$, which has structures VII, VIII, and IX.

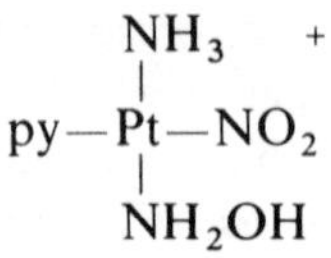

$$
\begin{array}{c}
NH_3 \quad + \\
| \\
py-Pt-NO_2 \\
| \\
NH_2OH
\end{array}
\qquad
\begin{array}{c}
NO_2 \quad + \\
| \\
py-Pt-NH_3 \\
| \\
NH_2OH
\end{array}
\qquad
\begin{array}{c}
py \quad + \\
| \\
O_2N-Pt-NH_3 \\
| \\
NH_2OH
\end{array}
$$

VII 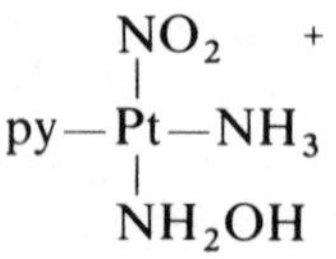   VIII 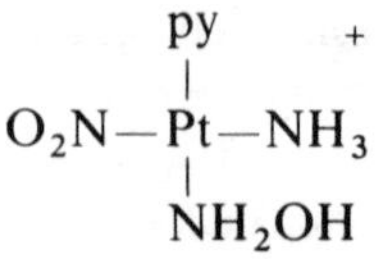   IX

Geometrical isomerism is also found in square planar systems containing unsymmetrical bidentate ligands, $[M(AB)_2]$. Glycinate ion, $NH_2CH_2COO^-$, is such a ligand; it coordinates with platinum(II) to form *cis*- and *trans*-$[Pt(gly)_2]$ having structures X and XI. It is not necessary for the attached ligand atoms to differ; all that is required is that the two halves of the chelate ring be different.

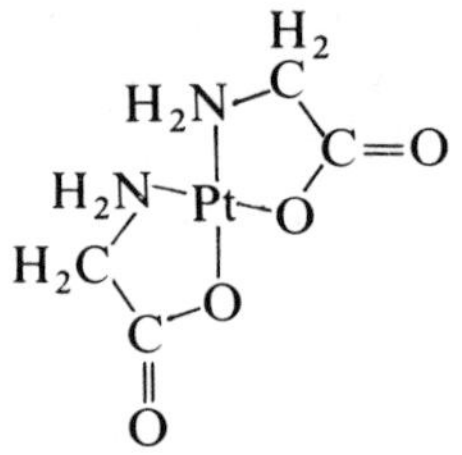

cis-diglycinatoplatinum(II)   *trans*-diglycinatoplatinum(II)

X   XI

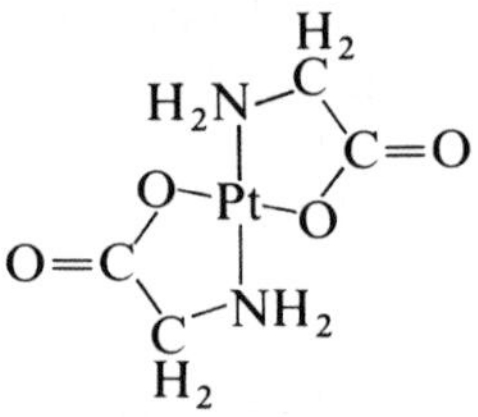

Geometrical isomerism in octahedral compounds is very closely related to that in square planar complexes. Among the most familiar examples of octahedral geometrical isomers are the violet (*cis*) and green (*trans*) forms of the tetraamminedichlorocobalt(III) and chromium(III) cations, which have structures XII and XIII. The largest number of geometrical isomers would exist for

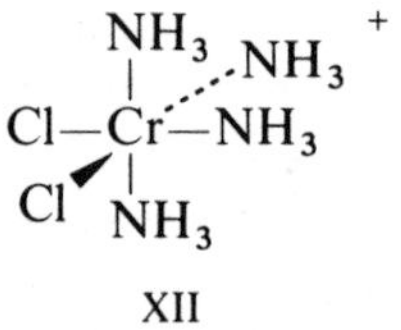

XII 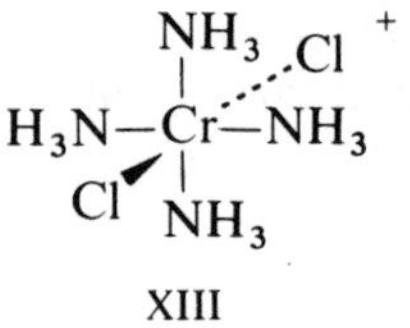   XIII

a complex of the type $[MABCDEF]$, wherein each ligand is different. Such a species can exist in fifteen different geometrical forms (each form would also have an optical isomer). Students may wish to draw all of the possible structures. The only compound of this type that has been prepared is $[Pt(NO_2)(Cl)(Br)(I)(py)(NH_3)]$. It was originally obtained in three different forms and no attempt was made to isolate all fifteen isomers.

Unsymmetrical bidentate ligands give rise to geometrical isomers in much the same way as was described earlier for square planar complexes. For example, the *cis*–*trans* isomers of triglycinatochromium(III) have the structures

XIV and XV.[1] Each of these complexes is optically active, as is discussed below. In [MA$_3$B$_3$] type complexes, such as XIV and XV, the *cis* isomer is called *facial*, since the like donor atoms are on the same face of the octahedron. The *trans* isomer is called *meri*dional.

XIV   XV

Optical isomerism

It has already been necessary to refer to the phenomenon of optical isomerism. A brief discussion is provided here, along with a few examples of optically-active metal complexes. Optical isomerism has been recognized for many years. The classical experiments in 1848 of Louis Pasteur, one of the most illustrious and humane of all men of science, showed that sodium ammonium tartrate exists in two different forms. Crystals of the two forms differ, and Pasteur was able to separate them by the laborious task of hand picking.

Aqueous solutions of the two isomers had the property of rotating a plane of polarized light (a beam of light vibrating in only one plane) either to the right or to the left. Because of this property the isomers are said to be *optically active* and are called *optical isomers*; one is designated as the *dextro* (+) isomer and the other as the *levo* (−) isomer. The extent of rotation of the plane of polarized light by the two isomers is exactly the same; however, the *dextro* isomer rotates the plane of light to the right, the *levo* isomer to the left. It follows that the rotations cancel each other in solutions containing equal concentrations of the two isomers. Such a (+), (−) mixture is called a *racemic mixture*. Since in solution it does not rotate a plane of polarized light, it is optically inactive.

The geometric relationship of most optical isomers is similar to that of the right and left hands, or feet, or gloves, or shoes. There is a rather subtle difference between the structures; the relative positions of the thumb and fingers on each hand are the same, yet the two hands are different. One is the mirror image of the other. An analogous situation must exist if a molecule or ion is to be optically active. In order for a molecule or ion to be optically active, it must not have a plane of symmetry; i.e., it should not be possible to divide the particle into two identical halves. The most basic test that can be applied in

---

[1] It is convenient to use abbreviations for chelating ligands in diagrams. In this text the presence of a chelating group is indicated by a curved line on which an abbreviation for the ligand is written. The type of atom bonded to the metal is also indicated.

attempting to decide whether a given structure will be optically active is to compare it with its mirror image. If the structure and its mirror image are different, the structure is said to be dissymetric and the substance will normally exhibit optical activity.

The (+) and (−) isomers of a given compound are called enantiomers, which mean "opposite forms". They have almost identical chemical and physical properties. They differ primarily in the direction in which they rotate a plane of polarized light. This property permits them to be readily detected and to be distinguished. A simple instrument known as a *polarimeter* is used for this purpose.

Note that the physiological effects of enantiomers are sometimes profoundly different. Thus the (−)-nicotine that occurs naturally in tobacco is much more toxic than the (+)-nicotine that is made in the laboratory. Specific effects such as these are attributed to dissymetric reaction sites in biological systems. Since enantiomers are so similar, and since in chemical reactions the two forms are produced in equal amounts, special techniques are required to separate the two. This separation process is called *resolution*. Some resolution methods are described in Section 4.10. Often, a single optical isomer will rearrange to give a racemic mixture; the process is called *racemization*.

A simple example of a dissymetric molecule is one with a tetrahedral structure wherein the central atom is surrounded by four different atoms or groups. There are many examples of such molecules among organic compounds. The structures of optical isomers may be represented by the amino acids, structures XVI and XVII. Tetrahedral metal complexes are generally

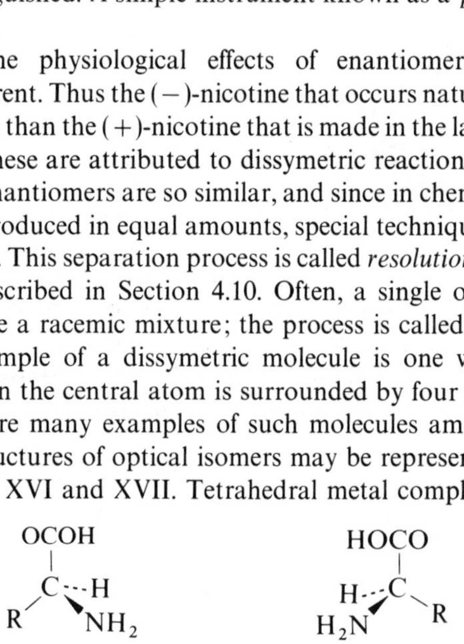

very reactive, which makes it extremely difficult to isolate them in isomeric forms. However, complexes containing two unsymmetrical bidentate ligands can be resolved into optically-active forms. The enantiomers of bis(benzoylacetonato)beryllium(II) have the structures XVIII and XIX. Note that four different groups around the central atom are not required for optical activity; the only requirement is that the molecule and its mirror image be different.

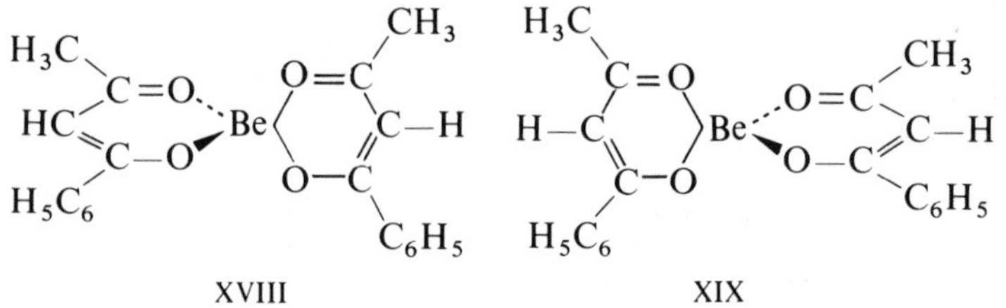

Square planar complexes are very seldom optically active. In most cases (for example, complexes of the type [MABCD]), the plane of the molecule is a plane of symmetry.

Unlike four-coordinated systems, six-coordinated complexes afford many examples of optical isomerism. These are very common among compounds or ions of the type $[M(AA)_3]$. For example, the optical isomers of trioxalatochromate(III) are complexes XX and XXI.

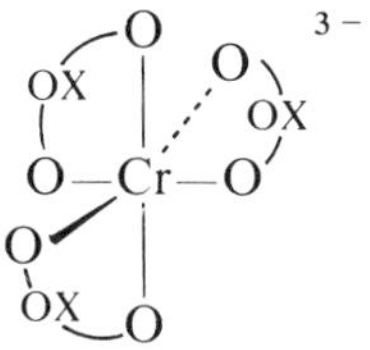

XX 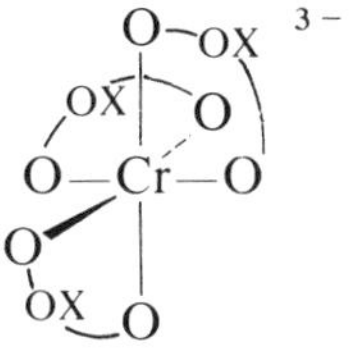   XXI

It was once claimed that optically-active compounds must contain carbon, but now at least three optically-active, purely inorganic complexes are known. One complex, prepared by Werner to show that carbon was not required, used the bridged complex, XXII, in which the dihydroxo complex XXIII is a

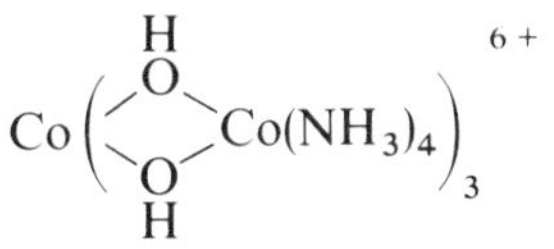

XXII 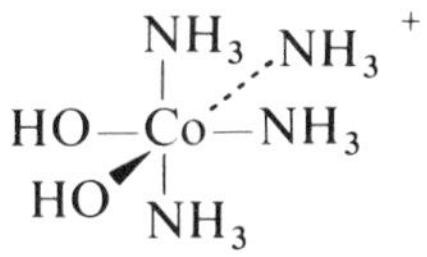   XXIII

bidentate ligand. The fact that complexes of the type $[M(AA)_3]$ can be resolved into optical isomers is good evidence that these complexes have the octahedral configuration. Neither a trigonal prism nor a planar structure would give rise to optical activity (Table 1.5).

Another very common type of optically-active complex has the general formula $[M(AA)_2X_2]$. In this system, it is important to note that the *trans* isomer has a plane of symmetry and cannot be optically active. Therefore, the *cis* structure for such a complex is conclusively demonstrated if the complex is shown to be optically active. This technique for proof of structure has often been used; the identity of the *cis* and *trans* isomers of the complex dichlorobis(ethylenediamine)rhodium(III), structures XXIV, XXV, and XXVI, was determined by this technique.

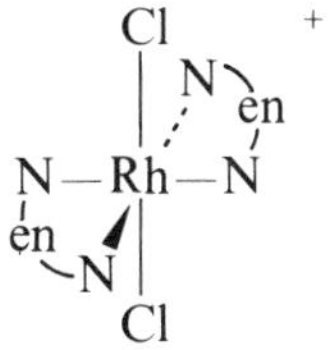

trans   *dextro-cis*   *levo-cis*

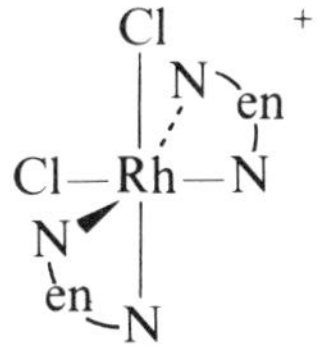

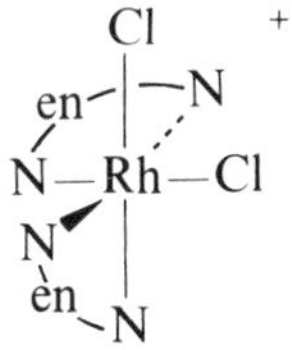

XXIV   XXV   XXVI

Multidentate ligands can also give rise to optical isomerism in metal complexes. One of the many such cases is that of (+)- and (−)-[Co(EDTA)]⁻, structures XXVII and XXVIII.

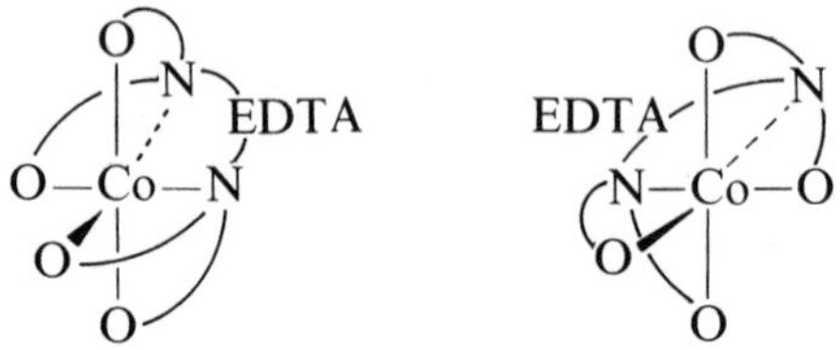

dextro	*levo*
XXVII	XXVIII

A complex containing six different ligands is dissymetric; each of its fifteen geometrical isomers should be resolvable into isomers. Thus for one form of $[Pt(NO_2)(Cl)(Br)(I)(py)(NH_3)]$ the optical isomers are structures XXIX and XXX.

dextro	*levo*
XXIX	XXX

In conclusion, note that the designation of an optical isomer as either *dextro* or *levo* is meaningful only if the wavelength of the light used is known. That an optical isomer may rotate the plane of polarized light to the right (*dextro*) at one wavelength but to the left at another is clearly shown in Figure 3.4. The mirror image isomer gives the mirror image curve. Such plots of optical rotation versus wavelength of light are called *rotatory dispersion curves*. They are more meaningful and useful than just the optical rotation at one wavelength. The absolute configuration of $(+)$-$[Co(en)_3]^{3+}$ was determined by means of X-ray diffraction studies. Using this as a standard it has been possible to assign absolute structures to other complexes by comparison of their rotatory dispersion curves.

Other types of isomerism

Several types of isomerism, other than geometrical and optical, are known for coordination compounds. These are often unique to this class of compound. Specific examples are given below to represent each type. In general, the nature of the isomerism is sufficiently obvious from the examples that no lengthy discussion is required.

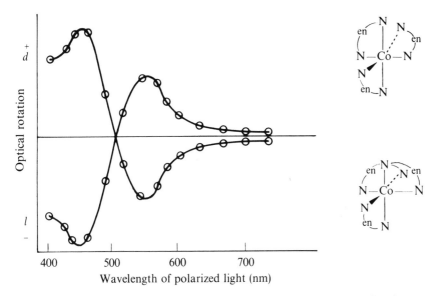

Figure 3.4 The rotatory dispersion curves and the structures for the optical isomers of $[Co(en)_3]^{3+}$.

Coordination isomerism

Compounds containing both cationic and anionic complexes are capable of forming isomers of the coordination isomer type whenever two different, random combinations are possible between and including the extremes $[MA_n][M'X_m]$ and $[M'A_m][MX_n]$. Some examples are

$[Co(NH_3)_6][Cr(C_2O_4)_3]$ and $[Cr(NH_3)_6][Co(C_2O_4)_3]$
$[Rh(en)_3][IrCl_6]$, $[RhCl_2(en)_2][IrCl_4(en)]$, and $[Ir(en)_3][RhCl_6]$
$[Pt(II)(NH_3)_4][Pt(IV)Cl_6]$ and $[Pt(IV)Cl_2(NH_3)_4][Pt(II)Cl_4]$
$[Cr(NH_3)_6][Cr(NCS)_6]$ and $[Cr(NCS)_2(NH_3)_4][Cr(NCS)_4(NH_3)_2]$

A special type of coordination isomerism is one that involves the different placement of ligands in a bridged complex. This is sometimes called *coordination position isomerism*, and an example is provided by the isomers

$$[(NH_3)_4Co\underset{\underset{H}{O}}{\overset{\overset{H}{O}}{<\,>}}Co(NH_3)_2Cl_2]SO_4$$

and

$$[Cl(NH_3)_3Co\underset{\underset{H}{O}}{\overset{\overset{H}{O}}{<\,>}}Co(NH_3)_3Cl]SO_4$$

Ionization isomerism

Ionization isomerism is the term used to describe isomers that yield different ions in solution. A classical example is the purple $[CoBr(NH_3)_5]SO_4$ and red $[CoSO_4(NH_3)_5]Br$, which give sulfate and bromide ions, respectively, in solution. Two sets of the many isomers of this type are

$$[Co(NCS)_2(en)_2]Cl \text{ and } [Co(NCS)Cl(en)_2]NCS$$
$$[PtBr(NH_3)_3]NO_2 \text{ and } [PtNO_2(NH_3)_3]Br$$

Very similar to these are the isomers resulting from replacement of a coordinated group by water of hydration. This type of isomerism is sometimes called *hydration isomerism*. The best known example is the trio of compounds $[Cr(H_2O)_6]Cl_3$, $[CrCl(H_2O)_5]Cl_2 \cdot H_2O$, and $[CrCl_2(H_2O)_4]Cl \cdot 2H_2O$, which contain six, five, and four coordinate water molecules, respectively. These isomers differ markedly in physical and chemical properties. Other isomers of the same type are

$$[CoCl(en)_2(H_2O)]Cl_2 \text{ and } [CoCl_2(en)_2]Cl \cdot H_2O$$
$$[CrCl_2(py)_2(H_2O)_2]Cl \text{ and } [CrCl_3(py)_2(H_2O)] \cdot H_2O$$

Linkage isomerism

Isomerism of the linkage type may result whenever a ligand has two different atoms available for coordination. The linkage between the metal and ligand in one isomer is through one ligand atom, and that of its isomer is through another. It has long been known that nitrite ion in cobalt(III) complexes can be attached either through the nitrogen, $Co-NO_2$ (nitro), or the oxygen, $Co-ONO$ (nitrito). The nitrito complexes of cobalt(III) are unstable and rearrange to form the more stable nitroisomers. Similar linkage isomers can be obtained in complexes of rhodium(III), iridium(III), and platinum(IV). Some examples of this type of isomerism are

$$[(NH_3)_5Co-NO_2]Cl_2 \text{ and } [(NH_3)_5Co-ONO]Cl_2$$
$$[(NH_3)_2(py)_2Co(-NO_2)_2]NO_3 \text{ and } [(NH_3)_2(py)_2Co(-ONO)_2]NO_3$$
$$[(NH_3)_5Ir-NO_2]Cl_2 \text{ and } [(NH_3)_5Ir-ONO]Cl_2$$

All ligands other than NO_2^- are written to the left of the metal in order to emphasize how the nitrite ion is bound to the metal.

Many other ligands are potentially capable of forming linkage isomers. Theoretically, all that is required is that two different atoms of the ligand contain an unshared electron pair. Thus the thiocyanate ion, $:N:::C:\overset{..}{\underset{..}{S}}:^-$, can attach itself to the metal either through the nitrogen, $M-NCS$, or the sulfur, $M-SCN$. Each type of attachment does occur, but generally to give only one form or the other in any particular system. Usually the first-row transition elements are attached through nitrogen, whereas the second- and third-row transition elements (in particular the platinum metals) are attached through sulfur. The following linkage isomers of this type have been prepared.

[(bipy)Pd($-$SCN)$_2$] and [(bipy)Pd($-$NCS)$_2$]
[(OC)$_5$Mn$-$SCN] and [(OC)$_5$Mn$-$NCS]
[(H$_2$O)$_5$Cr$-$SCN]$^{2+}$ and [(H$_2$O)$_5$Cr$-$NCS]$^{2+}$

Linkage isomers of CN$^-$ and of S$_2$O$_3^{2-}$ metal complexes have also been reported.

Conformational isomerism

As was stated earlier, four-coordinated complexes have either a tetrahedral or a square planar structure. The geometry observed in a particular system depends on the metal and also on the ligands. Thus complexes of beryllium(II) are always tetrahedral, whereas nickel(II) forms tetrahedral [NiBr$_4$]$^{2-}$ and square planar [Ni(CN)$_4$]$^{2-}$. For certain ligands the stabilities of the two structures may not differ greatly; in such cases both forms may be obtained and these are called *conformational isomers*. Examples of such isomers for nickel(II) and cobalt(II) are

(L = (C$_6$H$_5$)$_2$(C$_6$H$_5$CH$_2$)P).

Problems

1. Predict the geometry of the following ions:

[PtCl$_6$]$^{2-}$ (diamagnetic); [NiF$_6$]$^{4-}$ (two unpaired electrons);
[Cr(H$_2$O)$_6$]$^{2+}$ (four unpaired electrons);
[V(H$_2$O)$_6$]$^{2+}$ (three unpaired electrons);
[FeCl$_4$]$^-$ (five unpaired electrons);
[PdCl$_4$]$^{2-}$ (diamagnetic).

2. Draw all of the possible isomers of the following complexes:
[CrBr$_2$(NH$_3$)$_4$]$^+$ [Be(gly)$_2$]
[Co(C$_2$O$_4$)$_2$(en)]$^-$ [Pt(en)(NO$_2$)$_2$Cl$_2$]
[Pt(gly)$_2$Cl$_2$]   [Co$_2$(OH)$_2$Cl$_2$(NH$_3$)$_6$]$^{2+}$
[Pt(gly)$_2$]   [Al(EDTA)]$^-$

3. (*a*) A complex [M(AB)$_2$] is known to be optically active. What does this indicate about the structure of the complex? (*b*) A complex of the type [MX$_2$(AA)$_2$] is known to be optically active. What does this indicate about the structure of the complex?

4. Would you expect VCl_4^-, $TiCl_4$, and $MnCl_4^-$ to have regular or distorted tetrahedral geometry? Explain your answer.

5. (*a*) How many geometric isomers can exist for octahedral complexes with these stoichiometries: MA_3B_3, MA_4BC, MA_3B_2C, and $MA_2B_2C_2$? (*b*) Will any of the geometric isomers be suitable for possible resolution into optical isomers?

6. Suggest a chemical procedure for determining which of the geometric isomers of $[CoCl_2(en)(NH_3)_2]^+$ is the completely *cis* form.

References

Wells, A. F. *Structural Inorganic Chemistry*, 4th edn (Clarendon, Oxford, 1975).
Muetterties, E. L. and Wright, C. M. "Molecular Polyhedra of High Coordination Number", *Q. Rev.*, **21**, 109 (1967).
Muetterties, E. L. "Metal Clusters", *Chem. Engng News*, 30 August, 28 (1982).
Cotton, F. A. and Chisholm, M. H. "Bonds between Metal Atoms", *Chem. Engng News*, 28 June, 40 (1982).

Chapter 4
Preparations and Reactions of Coordination Compounds

The preparation of compounds has always been a most important part of chemistry. Certainly, research in chemical industry is largely oriented toward the synthesis of new and useful materials. The chemist is very interested in preparing new compounds, because it is an excellent way of increasing our knowledge of chemistry. Chapter 1 relates how the synthesis of the first coordination compounds led to the development of concepts and theories that are of considerable value today. The preparation of ferrocene $(Fe(C_5H_5)_2)$ is another example of a synthesis that has led to a tremendous amount of research effort in both synthetic and theoretical chemistry. This discovery of ferrocene marked the beginning of vigorous research activity in transition-metal organometallic chemistry and in homogeneous catalysis (see Chapter 7).

Several different but related experimental methods can be used to prepare metal complexes. Some of these are described in this chapter. The choice of method depends upon the system in question, and not all methods are applicable to the synthesis of a particular compound. Finding a reaction that produces the desired compound in good yield is only the beginning. The next step is to find a suitable way to isolate the product from its reaction mixture. Crystallization has been most frequently used. The more commonly used techniques include:

(1) Evaporate the solvent and cool the more concentrated reaction mixture in an ice–salt bath. Adding a seed crystal of the desired compound and scratching the inside of the beaker below the liquid surface often helps to induce crystallization.

(2) Slowly add a solvent that is miscible with the solvent of the reaction mixture but that does not dissolve the desired compound. The techniques of cooling, seeding, and scratching may be necessary to cause the product to precipitate from the mixed solvent.

(3) If the desired complex is a cation, it may be isolated by the addition of an appropriate anion with which it forms an insoluble salt. A suitable cation may be added to the reaction mixture to precipitate an anionic complex. Often a large ion of equal but opposite charge to that of the complex facilitates its

isolation. For example, the presence of $Ni(CN)_5^{3-}$ in solution had been demonstrated, but all attempts to isolate it failed until a large trivalent cation was used, to give $[Cr(en)_3][Ni(CN)_5] \cdot 1.5H_2O$.

Chromatographic processes can be very effective in the separation and purification of complexes. Techniques such as distillation and sublimation can be used to separate complexes which are volatile.

4.1 Substitution reactions in aqueous solution

The substitution reaction in aqueous solution is the most common method used for the synthesis of metal complexes. The method involves a reaction between a metal salt in water solution and a coordinating agent. For example, the complex $[Cu(NH_3)_4]SO_4$ is readily prepared by the reaction (1) between

$$\underset{\text{blue}}{[Cu(H_2O)_4]^{2+}} + 4NH_3 \rightarrow \underset{\text{dark blue}}{[Cu(NH_3)_4]^{2+}} + 4H_2O \qquad (1)$$

an aqueous solution of $CuSO_4$ and excess NH_3. That the coordinated water is instantly replaced by ammonia at room temperature as indicated by the change in color from light blue to dark blue. The dark-blue salt crystallizes from the reaction mixture upon the addition of ethanol.

Substitution reactions of metal complexes can also be fairly slow and, for such systems, more drastic experimental conditions are required. To prepare $K_3[Rh(C_2O_4)_3]$, one must boil a concentrated aqueous solution of $K_3[RhCl_6]$ and $K_2C_2O_4$ for 2 hours and then evaporate it until the product crystallizes from the solution, reaction (2).

$$\underset{\text{wine red}}{K_3[RhCl_6]} + 3K_2C_2O_4 \xrightarrow[\underset{100°C}{}]{\overset{H_2O}{\overset{2h}{}}} \underset{\text{yellow}}{K_3[Rh(C_2O_4)_3]} + 6KCl \qquad (2)$$

It is also possible that more than one type of ligand will be replaced during a reaction. Thus $[Co(en)_3]Cl_3$ may be prepared by reaction (3). Because this reaction is rather slow at room temperature, it is carried out on a steam bath.

$$\underset{\text{purple}}{[CoCl(NH_3)_5]Cl_2} + 3en \rightarrow \underset{\text{orange}}{[Co(en)_3]Cl_3} + 5NH_3 \qquad (3)$$

The examples given above are for the preparation of complexes containing only the entering ligand. Such complexes are the easiest to prepare because an excess of coordinating agent may be used to force the equilibrium toward the completely substituted complex. Theoretically, it should be possible to obtain the intermediate mixed complexes because substitution reactions proceed in a stepwise fashion (Section 5.1). In practice, however, it is often very difficult to isolate the desired mixed complex directly from the reaction mixture. Some successful syntheses of mixed compounds have been achieved by limiting the concentration of a potential ligand. The complex $[Ni(phen)_2(H_2O)_2]Br_2$ can be isolated from a reaction mixture containing two equivalents of phen to one of $NiBr_2$. Similarly, the compound diammineethylenediamineplatinum(II) chloride can be prepared by reactions (4) and (5). Reaction (4) is successful primarily because the dichloro product is nonionic and separates from the aqueous reaction mixture as it is formed.

$$K_2[\underset{red}{PtCl_4}] + en \rightarrow [\underset{yellow}{PtCl_2(en)}] + 2KCl \tag{4}$$

$$[\underset{yellow}{PtCl_2(en)}] + 2NH_3 \rightarrow [\underset{colorless}{Pt(NH_3)_2(en)}]Cl_2 \tag{5}$$

Since the nitrogen molecule is isoelectronic with CO, CN^-, and NO^+, all of which form a variety of stable molecules, then N_2 should also have some tendency to coordinate with metals. Although coordination chemists had long tried to make N_2 complexes, the first successful synthesis (of $[Ru(NH_3)_5(N_2)]Br_2$) was not achieved until 1965. This important discovery was immediately followed by intensive research, and many different dinitrogen–metal (N_2–M) complexes were prepared. In several cases, it was found that simple ligand substitution (6) can generate the desired product.

$$[Ru(NH_3)_5H_2O]^{2+} + N_2 \xrightarrow[25°C]{H_2O} [Ru(NH_3)_5(N_2)]^{2+} + H_2O \tag{6}$$

Chemists are very excited about these N_2 complexes because the fixation of dinitrogen (the conversion of molecular nitrogen into other nitrogen compounds) is commercially very important. Since the discovery of dinitrogen–metal complexes, several research groups have demonstrated that dinitrogen can be converted into ammonia under ambient laboratory conditions. Unfortunately, the reactions known today do not provide a less expensive method for making ammonia than the high-temperature, high-pressure Haber process, which has been used for decades. It has long been known that certain bacteria associated with clover, peas, and other plants do fix nitrogen in nature and that transition metals are involved in these processes: the metallo-enzyme present is nitrogenase, a protein containing Mo and Fe. Although the mechanism of nitrogen fixation is not known, it appears that M–N_2 complexes are involved.

4.2 Substitution reactions in nonaqueous solvents

Reactions in solvents other than water had not been used extensively for the preparation of metal complexes until recently. Two of the chief reasons why it is sometimes necessary to use solvents other than water are that either the metal ion has a large affinity for water or the ligand is insoluble in water. A few common ions that have a large affinity for water, and thus form strong metal–oxygen bonds, are aluminum(III), iron(III), and chromium(III). The addition of basic ligands to aqueous solutions of these metal ions generally results in the formation of a gelatinous hydroxide precipitate rather than a complex containing the added ligands. The metal–oxygen bonds remain intact, but oxygen–hydrogen bonds are broken; the hydrated metal ions behave as protonic acids.

The aqueous reaction between a chromium(III) salt and ethylenediamine is shown by equation (7). If, instead, an anhydrous chromium salt and a non-

$$[Cr(H_2O)_6]^{3+} + 3en \xrightarrow{H_2O} [Cr(OH)_3(H_2O)_3]\downarrow + 3enH^+ \qquad (7)$$
$$\underset{\text{violet}}{} \qquad\qquad\qquad \underset{\text{green}}{}$$

aqueous solvent are used, the reaction proceeds smoothly to yield the complex $[Cr(en)_3]^{3+}$, equation (8). Although many amminechromium(III) complexes

$$CrCl_3 + 3en \xrightarrow{\text{ether}} [Cr(en)_3]Cl_3 \qquad (8)$$
$$\underset{\text{purple}}{} \qquad\qquad \underset{\text{yellow}}{}$$

are known, very few of them are ever prepared directly by reaction in water solution. One solvent that has been used rather extensively is dimethyl-formamide (DMF), $(CH_3)_2NCHO$. By using this solvent it was possible to prepare *cis*-$[CrCl_2(en)_2]Cl$ in good yield by direct reaction (9).

$$[CrCl_3(DMF)_3] + 2en \xrightarrow{\text{DMF}} cis\text{-}[CrCl_2(en)_2]Cl \qquad (9)$$
$$\underset{\text{blue-gray}}{} \qquad\qquad\qquad \underset{\text{violet}}{}$$

In some cases, a nonaqueous solvent is required because the ligand is not water soluble. Often it suffices to dissolve the ligand in a water-miscible solvent and then to add this solution to a concentrated water solution of the metal ion. Metal complexes of bipy and phen are generally prepared by this way. Thus the addition of an alcoholic solution of bipy to an aqueous solution of $FeCl_2$ readily yields the complex $[Fe(bipy)_3]Cl_2$, reaction (10).

$$[Fe(H_2O)_6]^{2+} + 3bipy \xrightarrow{H_2O-C_2H_5OH} [Fe(bipy)_3]^{2+} + 6H_2O \qquad (10)$$
$$\underset{\text{colorless}}{} \qquad\qquad\qquad\qquad \underset{\text{intense red}}{}$$

4.3 Substitution reactions in the absence of solvent

The direct reaction between an anhydrous salt and a liquid ligand can be used to prepare metal complexes. In many cases, the liquid ligand present in very large excess also serves as a solvent for the reaction mixture. One method, which can be used to synthesize metal ammines, involves the addition of a metal salt to liquid ammonia followed by evaporation to dryness. Evaporation occurs readily at room temperature, because ammonia boils at $-33°C$. The dry residue obtained is essentially the pure metal ammine. For example, $[Ni(NH_3)_6]Cl_2$ can be prepared in this way by reaction (11).

$$NiCl_2 + 6NH_3(\text{liquid}) \rightarrow [Ni(NH_3)_6]Cl_2 \qquad (11)$$
$$\underset{\text{yellow}}{} \qquad\qquad\qquad\qquad \underset{\text{violet}}{}$$

Often, this is not the method of choice, because aqueous ammonia is more convenient to use and generally gives the same result. However, in some cases, as in the preparation of $[Cr(NH_3)_6]Cl_3$, it is necessary to use liquid ammonia to avoid the formation of $Cr(OH)_3$ (see equation (7)).

One method used for the preparation of $[Pt(en)_2]Cl_2$ or $[Pt(en)_3]Cl_4$ is the direct reaction between ethylenediamine and $PtCl_2$ or $PtCl_4$, respectively. The technique is to add slowly the solid platinum salts to the liquid ethylenediamine. This addition is accompanied by a vigorous evolution of heat, which is to be expected whenever a strong acid is added to a strong base. Recall (Section 2.1) that in terms of the Lewis definition of acids and bases, the formation of coordination compounds involves an acid–base reaction. In this particular case, the platinum ions are the acids and ethylenediamine is the base. Metal dimethylsulfoxide complexes have been prepared and characterized. One method used to prepare some of these complexes is a direct reaction (12) in the absence of any added solvent.

$$Co(ClO_4)_2 + 6(CH_3)_2SO \rightarrow [Co\{(CH_3)_2SO\}_6](ClO_4)_2 \qquad (12)$$
$$\text{pink} \qquad\qquad\qquad\qquad \text{pink}$$

4.4 Thermal dissociation of solid complexes

Thermal dissociation amounts to a substitution reaction in the solid state. At some elevated temperature, volatile coordinated ligands are lost and their place in the coordination sphere is taken by the anions of the complex. A familiar example, but one that is seldom considered in this way, is the loss of water by $CuSO_4 \cdot 5H_2O$ when it is heated. The blue hydrate yields the almost white anhydrous sulfate by reaction (13). The replacement of water ligands by

$$[Cu(H_2O)_4]SO_4 \cdot H_2O \xrightarrow{\text{heat}} [CuSO_4] + 5H_2O\uparrow \qquad (13)$$
$$\text{blue} \qquad\qquad\qquad \text{colorless}$$

sulfate ions is responsible for the change in color. The hydrated copper(II) ion absorbs light near the infrared end of the visible spectrum; this is responsible for its blue color. Since the crystal field splitting due to sulfate ion is less than that of water, copper(II) ion in a sulfate ion environment absorbs light of a longer wavelength. This places the anhydrous copper sulfate absorption in the infrared; hence the compound has no absorption in the visible spectrum and is colorless. The situation described on page 2 for invisible ink is another illustration of a solid-state reaction.

At elevated temperatures, coordinated water can often be liberated from solid aquoamminemetal complexes. It is sometimes convenient to use this method to prepare halogenoamminemetal compounds (14).

$$[RhH_2O(NH_3)_5]I_3 \xrightarrow{100°C} [RhI(NH_3)_5]I_2 + H_2O\uparrow \qquad (14)$$
$$\text{colorless} \qquad\qquad \text{yellow}$$

Just as water can be expelled from solid aquo complexes, ammonia and amines can sometimes be liberated from metal ammines. This procedure is used to prepare acidoamminemetal[1] complexes. This is a general method for the synthesis of compounds of the type *trans*-$[PtA_2X_2]$. The reaction yields the *trans* isomer, as described in Section 4.8. The most common example of this is the preparation of *trans*-$[Pt(NH_3)_2Cl_2]$ by the thermal evolution of ammonia, reaction (15). The corresponding reaction for the analogous

$$[Pt(NH_3)_4]Cl_2 \xrightarrow{\;250°C\;} trans\text{-}[PtCl_2(NH_3)_2] + 2NH_3\uparrow \qquad (15)$$
$$\underset{white}{} \qquad\qquad \underset{yellow}{}$$

pyridine system takes place at a temperature approximately 100°C lower.

4.5 Photochemical synthesis

We have seen that compounds can be thermally dissociated to produce new complexes. Another way of imparting energy to a molecule to induce reaction is to shine on it light of an appropriate wavelength. There are many important applications of photochemistry in organometallic chemistry but it has been used less frequently with metal ammine and other classical complexes.

A classical example of photochemical synthesis of a metal ammine complex is the conversion of a stable nitro to unstable nitrito complex, reaction (16).

$$[(NH_3)_5Co-NO_2]X_2(s) \xrightarrow{\;h\nu\;} [(NH_3)_5Co-ONO]X_2(s) \qquad (16)$$

Metal oxalates often react photochemically to produce CO_2 and the metal in a lower oxidation state. This reaction has recently been used to generate a coordinatively-unsaturated platinum which then functions as an efficient homogeneous catalyst, reaction (17).

$$[Pt(C_2O_4)(PR_3)_2] \xrightarrow{\;h\nu\;} [Pt(PR_3)_2] + 2CO_2 \qquad (17)$$

In the presence of other ligands, this reaction can be used to prepare Pt(O) compounds of the type $PtL_2(PR_3)_2$.

Irradiation of a metal carbonyl with ultraviolet light often results in the loss of CO. This approach can be used to replace CO from compounds which do not react or react very slowly, equation (18).

$$W(CO)_6 + PPh_3 \xrightarrow[h\nu,\,fast]{-CO} \overset{\Delta,\,slow}{} W(CO)_5PPh_3 \qquad (18)$$

---

[1] Acido is used as a general term referring to anionic ligands. Likewise, the use of ammine with reference to a general class of compounds is not specific for ammonia but includes other amines. Thus $[Co(NH_3)_4Br_2]^+$, $[Cr(en)(C_2O_4)_2]^-$, and $[Pt(py)_2(NO_2)_2]$ are all examples of acidoammine complexes.

4.6   Oxidation–reduction reactions

The preparation of many metal complexes often involves an accompanying oxidation–reduction reaction. For the thousands of cobalt(III) complexes that have been prepared, the starting material was almost always some cobalt(II) salt. This is because the usual oxidation state of cobalt in its simple salts is 2. The oxidation state of 3 becomes the stable form only when cobalt is coordinated to certain types of ligands (Section 5.2). Furthermore, it is convenient to start with salts of cobalt(II) because cobalt(II) complexes undergo substitution reactions very rapidly, whereas reactions of cobalt(III) complexes are very slow (Section 6.4). The preparation of cobalt(III) complexes, therefore, proceeds by a fast reaction between cobalt(II) and the ligand to form a cobalt(II) complex which is then oxidized to the corresponding cobalt(III) complex. For example, reaction (19) is presumed to involve first the formation of $[Co(NH_3)_6]^{2+}$

$$4[Co(H_2O)_6]Cl_2 + 4NH_4Cl + 20NH_3 + O_2 \rightarrow 4[Co(NH_3)_6]Cl_3 + 26H_2O$$
pink   orange

(19)

reaction (20), followed by its oxidation, reaction (21).

$$[Co(H_2O)_6]Cl_2 + 6NH_3 \rightarrow [Co(NH_3)_6]Cl_2 + 6H_2O \qquad (20)$$
pink   rose

$$4[Co(NH_3)_6]Cl_2 + 4NH_4Cl + O_2 \rightarrow 4[Co(NH_3)_6]Cl_3 + 4NH_3 + 2H_2O$$
rose   orange

(21)

Although air oxidation is commonly used in the synthesis of cobalt(III) complexes, other oxidizing agents can be employed. Many oxidizing agents are able to oxidize cobalt(II) to cobalt(III) in the presence of suitable ligands, but only a few are convenient to use. Oxidizing agents such as potassium permanganate and potassium dichromate introduce to the reaction mixture ions that are not easily separated from the desired product. Oxidizing agents such as oxygen and hydrogen peroxide do not introduce foreign metal ions to the reaction mixture. Another type of suitable oxidizing agent is one whose reduction product is insoluble and can be removed by filtration. This is true of PbO_2, which is reduced to Pb^{2+} and can be removed as insoluble $PbCl_2$. Similarly, SeO_2 yields insoluble Se. For a more complete discussion of oxidation–reduction reactions of metal complexes see Section 6.8.

Less common than preparation of complexes by the oxidation of the central metal ion is preparation by reduction to complexes of the metal ion in a lower oxidation state. One reason that the latter has not been extensively applied is that the resulting compounds are often so sensitive to oxidation that they must be handled in an inert, oxygen- and moisture-free atmosphere. However, with special precautions, it is possible to prepare many interesting complexes in which the central metal ion has an unusually low oxidation state. Reductions in liquid ammonia have been useful for this purpose, as is illustrated by reaction (22). The oxidation state of nickel in this compound is zero. The

$$K_2[Ni(CN)_4] + 2K \xrightarrow[\text{ammonia}]{\text{liquid}} K_4[Ni(CN)_4] \qquad (22)$$

$$\text{yellow} \qquad\qquad\qquad\qquad\qquad \text{yellow}$$

compound is readily oxidized in air, and it reduces water with the liberation of hydrogen. It is even possible to reduce the central metal ion of a complex to a negative oxidation state. Iron has an oxidation state of -2 in $K_2[Fe(CO)_4]$, which is prepared by reaction (23). The salt is stable in aqueous

$$[Fe(CO)_5] + 4KOH \rightarrow K_2[Fe(CO)_4] + K_2CO_3 + 2H_2O \qquad (23)$$
$$\text{yellow} \qquad\qquad\qquad \text{colorless}$$

alkaline solution, but it is very sensitive to air oxidation. Another example of a complex containing a metal ion in a negative oxidation state is $[V(bipy)_3]^-$, which is prepared by the reduction of $[V(bipy)_3]^{3+}$. Note that in most cases of unusually low oxidation states, the EAN (Section 2.3) of the metal ion is the same as that of the next noble gas.

4.7   Catalysis

In systems that react slowly, it is often necessary to employ elevated temperatures and long reaction times in order to prepare desired coordination compounds. Alternatively, a catalyst may be used to increase the speed of a reaction. Catalysis has been used successfully in a few instances to prepare metal complexes. There are two types of catalysis: *heterogeneous catalysis* takes place when the catalyst is in a different phase than that of the reactants; *homogeneous catalysis* occurs when the catalyst and reacting materials are in the same phase. Examples of the use of heterogeneous and homogeneous catalysis in the synthesis of metal complexes are noted below.

The best known example of heterogeneous catalysis in such systems is the preparation of $[Co(NH_3)_6]Cl_3$. The air oxidation of a reaction mixture of aqueous cobalt(II) chloride, excess ammonia, and ammonium chloride, followed by acidification with excess hydrochloric acid, yields largely $[Co(NH_3)_5Cl]Cl_2$, reaction (24). Under the same conditions, in the presence

$$[Co(H_2O)_6]Cl_2 \xrightarrow[\text{NH}_4\text{Cl}]{\text{NH}_3-\text{H}_2\text{O}-\text{O}_2} \xrightarrow{\text{HCl}} [CoCl(NH_3)_5]Cl_2 \qquad (24)$$
$$\text{pink} \qquad\qquad\qquad\qquad\qquad\qquad \text{purple}$$

of charcoal the product is almost exclusively $[Co(NH_3)_6]Cl_3$, reaction (25).

$$[Co(H_2O)_6]Cl_2 \xrightarrow[\text{NH}_4\text{Cl-charcoal}]{\text{NH}_3-\text{H}_2\text{O}-\text{O}_2} \xrightarrow{\text{HCl}} [Co(NH_3)_6]Cl_3 \qquad (25)$$
$$\text{pink} \qquad\qquad\qquad\qquad\qquad\qquad\quad \text{orange}$$

Why is the oxidation product (24) not the hexaammine in the absence of charcoal? The air oxidation of $[Co(NH_3)_6]^{2+}$ seems to proceed through a bridged intermediate. A molecule of oxygen may add to two of the reactive

$[Co(NH_3)_6]^{2+}$ cations to form a peroxo-bridged cobalt(III) species, reaction (26).

$$2[Co(NH_3)_6]^{2+} + O_2 \rightarrow [(NH_3)_5Co-O-O-Co(NH_3)_5]^{4+} + 2NH_3$$
orange   pink   (26)

Bridged complexes of this type are known. The bridged system must then react with ammonia to form $[Co(NH_3)_6]^{3+}$. This reaction is very slow; however, it occurs rapidly in the presence of decolorizing charcoal, reaction (27).

$$[(NH_3)_5Co-O-O-Co(NH_3)_5]^{4+} \xrightarrow[\text{C. fast}]{\text{very slow}}_{NH_3} [Co(NH_3)_6]^{3+} \quad (27)$$
pink   yellow

The product $[CoCl(NH_3)_5]Cl_2$ is presumably formed from the reaction of HCl with $[(NH_3)_5Co-O-O-Co(NH_3)_5]^{4+}$, reaction (28).

$$[(NH_3)_5Co-O-O-Co(NH_3)_5]^{4+} \xrightarrow{HCl} [(NH_3)_5Co-OH_2]^{3+} \xrightarrow{HCl}$$
pink   pink

$$[CoCl(NH_3)_5]^{2+} \quad (28)$$
purple

Homogeneous catalysis is observed in the reactions of several platinum(IV) complexes. Such complexes generally react extremely slowly, but in the presence of catalytic amounts of platinum(II), reaction occurs rapidly. Platinum(II) catalysis has been used successfully to prepare new compounds of platinum(IV) as well as complexes previously obtained by other methods. The complex, *trans*-$[PtCl(SCN)(NH_3)_4]^{2+}$, has only been made by this method of platinum(II) catalysis, reaction (29).

$$\textit{trans-}[PtCl_2(NH_3)_4]^{2+} + SCN^- \xrightarrow{[Pt(NH_3)_4]^{2+}} \textit{trans-}[PtCl(SCN)(NH_3)_4]^{2+} + Cl^-$$
yellow   orange   (29)

Platinum(II) catalysis in these systems is believed to proceed by a mechanism that involves an activated bridged complex and a two-electron redox reaction. Such a process is represented by the reaction schemes (30) to (33). In reaction

$$\begin{bmatrix} (NH_3)_4 \\ Pt \end{bmatrix}^{2+} + SCN \rightleftharpoons \begin{bmatrix} (NH_3)_4 \\ Pt-SCN \end{bmatrix}^+ \quad (30)$$

$$\begin{bmatrix} (NH_3)_4 \\ Cl-Pt-Cl \end{bmatrix}^{2+} + \begin{bmatrix} (NH_3)_4 \\ Pt-SCN \end{bmatrix}^+ \rightleftharpoons \begin{bmatrix} (NH_3)_4 \quad (NH_3)_4 \\ Cl-Pt-Cl-Pt-SCN \end{bmatrix}^{3+} \quad (31)$$

$$\begin{bmatrix} (NH_3)_4 \quad (NH_3)_4 \\ Cl-Pt-Cl-Pt-SCN \end{bmatrix}^{3+} \rightleftharpoons \begin{bmatrix} (NH_3)_4 \\ Cl-Pt \end{bmatrix}^+ + \begin{bmatrix} (NH_3)_4 \\ Cl-Pt-SCN \end{bmatrix}^{2+} \quad (32)$$

$$\left[\begin{array}{c}(NH_3)_4\\Cl-Pt\end{array}\right]^+ \rightleftharpoons \left[\begin{array}{c}(NH_3)_4\\Pt\end{array}\right]^{2+} + Cl^- \tag{33}$$

(30), the catalyst $[Pt(NH_3)_4]^{2+}$ is associated weakly with SCN^-, which is present in large excess. Recall that evidence was cited in Section 3.1 for the coordination of a fifth and sixth group above and below the square plane of a four-coordinated planar complex.

4.8   Substitution reactions without metal–ligand bond cleavage

Some metal complexes form without the breakage of a metal–ligand bond. In the preparation of $[CoOH_2(NH_3)_5]^{3+}$ salts from $[CoCO_3(NH_3)_5]^+$, CO_2 is produced by cleavage of a carbon–oxygen bond which leaves the metal–oxygen bond intact, reaction (34). This can be demonstrated by running the

$$[(NH_3)_5Co-O\vdots CO_2]^+ + 2H^+ \rightarrow [(NH_3)_5Co-OH_2]^{3+} + CO_2 \quad (34)$$
$$\text{pink} \qquad\qquad\qquad\qquad\qquad \text{pink}$$

reaction in ^{18}O-labeled water. Both products contain oxygen with normal isotopic distribution. This indicates that solvent water does not contribute oxygen to the products, and hence that the oxygen must come from the reactants. A simple but not conclusive piece of evidence for the preservation of the $Co-O$ bond is the fact that the reaction is complete soon after the acidification of the carbonato compound. Since reactions that involve the breaking of $Co-O$ bonds in a variety of compounds are slow, the fast reaction suggests that another mechanism is involved in this case.

Reactions similar to (34) are believed to be fairly general, and they have been used to prepare aquo complexes from the corresponding carbonato complexes. Other systems such as $[(NH_3)_5Co-OSO_2]^+$ and $[(NH_3)_5Co-ONO]^{2+}$ may react in a similar fashion to produce $[(NH_3)_5Co-OH_2]^{3+}$ with the liberation of SO_2 and NO, respectively.

The reverse of these processes may also be expected to occur. Reaction (35) has been studied in detail, and it is found to occur without $Co-O$ bond

$$[(NH_3)_5Co-^{18}OH]^{2+} + N_2O_3 \rightarrow [(NH_3)_5Co-^{18}ONO]^{2+} + HNO_2 (35)$$
$$\text{pink} \qquad\qquad\qquad\qquad \text{pink}$$

cleavage. The best evidence for retention of the $Co-O$ bond is the observation that when $[Co(NH_3)_5OH]^{2+}$ was labeled with ^{18}O, the product $[Co(NH_3)_5{}^{18}ONO]^{2+}$ contained 99.4 per cent of the ^{18}O originally present in the starting material. It is reasonable to expect a similar behavior for reactions of hydroxo complexes with other acid anhydrides, for example, CO_2 and SO_2. Indeed, $[Co(NH_3)_5CO_3]^+$ can be prepared by the reaction of $[Co(NH_3)_5OH]^{2+}$ with CO_2.

A variety of other reactions occur without metal–ligand bond cleavage. Reactions of this type have been used to convert nitrogen-containing ligands to ammonia. Examples include the oxidation of N-bonded thiocyanate, reaction (36), and the reduction of N-bonded nitrite, reaction (37).

$$[(NH_3)_5Co-NCS]^{2+} \xrightarrow[\text{H}_2\text{O}]{\text{H}_2\text{O}_2} [(NH_3)_5Co-NH_3]^{3+} \qquad (36)$$
$$\text{orange} \qquad\qquad\qquad \text{orange}$$

$$[(NH_3)_3Pt-NO_2]^+ \xrightarrow[\text{HCl}-\text{H}_2\text{O}]{\text{Zn}} [(NH_3)_3Pt-NH_3]^{2+} \qquad (37)$$
$$\text{white} \qquad\qquad\qquad\qquad \text{white}$$

Coordinated ligands often exhibit their own characteristic reactions. For example, the hydrolysis (38) and fluorination (39) of PCl_3 can be accomplished while PCl_3 itself functions as a ligand.

$$[Cl_2Pt(PCl_3)_2] + 6H_2O \rightarrow [Cl_2Pt(P(OH)_3)_2] + 6HCl \qquad (38)$$
$$\text{yellow} \qquad\qquad\qquad \text{yellow}$$

$$[Ni(PCl_3)_4] + 4SbF_3 \rightarrow [Ni(PF_3)_4] + 4SbCl_3 \qquad (39)$$
$$\text{colorless} \qquad\qquad \text{colorless}$$

A type of reaction of considerable current interest involves the addition to, or substitution on, organic molecules coordinated to metal ions. The reactivity of molecules is frequently changed by coordination and thus novel reactions are possible. Moreover, metal ions may bind ligands in a specific configuration which will allow reactions with improbable geometric requirements to occur. For example, $[Co(en)_3]^{3+}$ reacts with ammonia and formaldehyde (HCHO) to give the product I. The three ethylenediamine rings become tied together

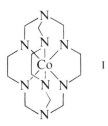

by $N(CH_2-)_3$ groups which form on both the top and bottom of the complex. The cobalt(III) is trapped inside the ligand which leads to several surprising properties. For example, an optical isomer of the complex can be reduced to the cobalt(II) complex and then reoxidized without racemization. (Racemization of other cobalt(II) complexes occurs very rapidly, as would be expected based on principles presented in Section 6.4.) In the formation of product I, the metal ion functions as a "template" on which the complicated organic molecule is constructed. In biological systems, metal ions also function as templates around which sterically-selective reactions occur.

4.9 *Trans* effect

The preparation of square planar compounds of platinum(II) has been put on a systematic basis by Russian workers, who observed that certain ligands cause the groups across from them in the square plane (*trans* position) to be replaced easily. Ligands that labilize the group *trans* to them are said to have a strong *trans*-directing influence (*trans* effect). The classic example of this is the preparation of the isomeric dichlorodiammineplatinum(II) complexes (40) and (41). In sequences (40) and (41), whenever a choice is possible, the group opposite to a Cl^- is replaced in preference to one *trans* to NH_3. On this basis Cl^- is said to have a greater *trans* effect than NH_3.

$$
\underset{\text{red}}{\overset{\displaystyle Cl}{\underset{\displaystyle Cl}{Cl-Pt-Cl}}}{}^{2-} \xrightarrow[NH_4^+]{NH_3} \underset{\text{orange}}{\overset{\displaystyle NH_3}{\underset{\displaystyle Cl}{Cl-Pt-Cl}}}{}^{-} \xrightarrow[NH_4^+]{NH_3} \underset{\text{yellow}}{\overset{\displaystyle NH_3}{\underset{\displaystyle Cl}{Cl-Pt-NH_3}}} \quad (40)
$$

$$
\underset{\text{colorless}}{\overset{\displaystyle NH_3}{\underset{\displaystyle NH_3}{H_3N-Pt-NH_3}}}{}^{2+} \xrightarrow{Cl^-} \underset{\text{yellow}}{\overset{\displaystyle NH_3}{\underset{\displaystyle NH_3}{Cl-Pt-NH_3}}}{}^{+} \xrightarrow{Cl^-} \underset{\text{yellow}}{\overset{\displaystyle NH_3}{\underset{\displaystyle NH_3}{Cl-Pt-Cl}}} \quad (41)
$$

After ligands are classified with respect to their *trans*-directing ability, the information can be used to synthesize specific desired compounds. The three isomers of $[Pt(NO_2)Cl(CH_3NH_2)NH_3]$ can be prepared by the reactions outlined in Figure 4.1. The success of these syntheses depends on the *trans* effect order: $NO_2^- > Cl^- > NH_3 \sim CH_3NH_2$. However, this information on the *trans* effect will not of itself permit one to propose such a reaction sequence. Another consideration is the stability of the various platinum–ligand bonds. The *trans* effect will explain the synthetic results in steps (*a*), (*c*), and (*f*); but the relative ease of replacement of chloride coordinated to platinum(II) accounts for steps (*b*), (*d*), and (*e*).

Extensive studies have shown that the *trans* effect of a variety of ligands decrease in the order:

$$CN^- \sim CO \sim C_2H_4 > PH_3 \sim SH_2 > NO_2^- > I^- > Br^- > Cl^-$$

$$> NH_3 \sim py > OH^- > H_2O$$

This information can now be used to design preparations of platinum(II) compounds or to predict their kinetic behavior. Investigations have demonstrated that the *trans* effect phenomenon may also be important in complexes of other metals.

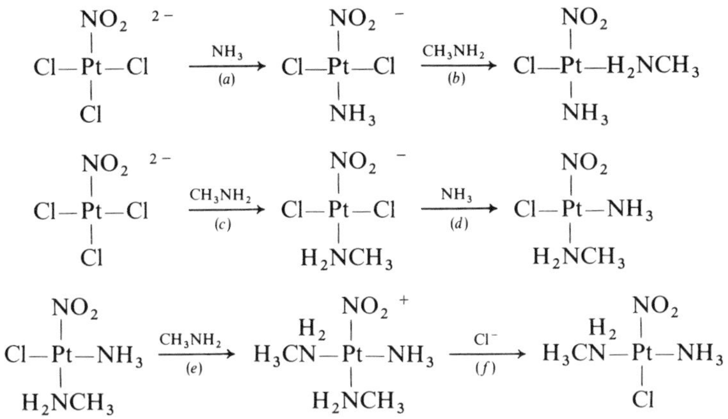

Figure 4.1 The preparation of the three isomers of $[Pt(NO_2)Cl(CH_3NH_2)(NH_3)]$.

4.10 Synthesis of *cis–trans* isomers

There are two approaches to the preparation of *cis–trans* isomers: preparation of a mixture of isomers (necessitating subsequent separation) and stereospecific synthesis that yields a single product. The second approach has been most successful for the synthesis of isomers of platinum(II) complexes by taking advantage of the *trans* effect (Section 4.8). Reactions of cobalt(III) complexes often yield a mixture of *cis–trans* isomers which must then be separated.

Complexes of platinum(IV) can readily be prepared as the *trans* isomer, whereas the *cis* form is much more difficult to obtain. The oxidation of square planar platinum(II) complexes generally yields the corresponding octahedral platinum(IV) complex. The preparation of *trans*-$[PtCl_2(NH_3)_4]^{2+}$ is achieved in this manner (42). Oxidizing agents such as Br_2 and H_2O_2 yield the

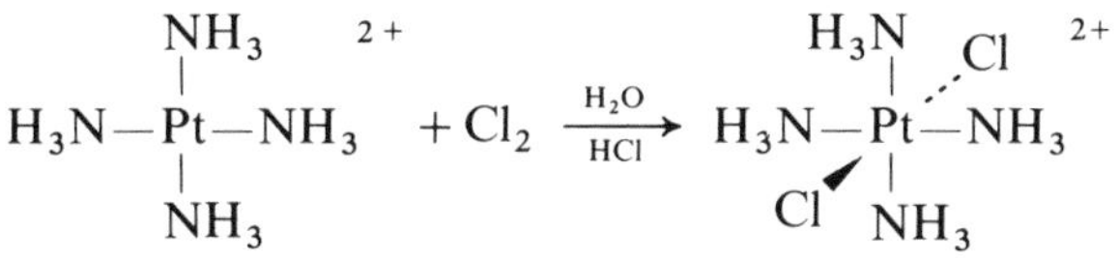

$$(42)$$

trans dibromo and dihydroxo complexes, respectively. These reactions are the earliest examples of oxidative addition reactions (see Section 7.2) which are now so important in organometallic chemistry. In some cases, this oxidation yields a *cis* product, for example,

$$[IrCl\{P(C_6H_5)_3\}_2CO] + Br_2 \rightarrow [IrClBr_2\{P(C_6H_5)_3\}_2CO]$$

Synthesis of *cis*-$[PtCl_2(NH_3)_4]Cl_2$ requires the more tedious approach given in reaction (43).

$$\begin{matrix} NH_3 \\ | \\ Cl-Pt-NH_3 \\ | \\ NH_3 \end{matrix}^{+} \xrightarrow[HCl]{Cl_2} \begin{matrix} H_3N \quad Cl \\ | \cdots | \\ Cl-Pt-NH_3 \\ \diagup | \\ Cl \quad NH_3 \end{matrix}^{+} \xrightarrow{NH_3} \begin{matrix} H_3N \quad NH_3 \\ | \cdots | \\ Cl-Pt-NH_3 \\ \diagup | \\ Cl \quad NH_3 \end{matrix}^{2+} \quad (43)$$

The replacement of one or more ligands in a particular isomer by other groups is sometimes used to prepare new compounds which have a desired stereochemistry. This is not a reliable method, however, because in many cases the stereochemistry of the starting material is not preserved during reaction.

The preparations of *cis-* and *trans-*$[PtCl_2(NH_3)_2]$, were discussed on page 74. In all of these, the product is largely the indicated isomer, but reactions frequently produce mixtures of isomers. These can be separated by means of fractional crystallization, ion-exchange chromatography, or other physical techniques. When KOH is added slowly to a refluxing wine-red aqueous solution of $RhCl_3 \cdot 3H_2O$ and $H_2NCH_2CH_2NH_2 \cdot 2HCl(en \cdot 2HCl)$, a clear yellow solution is obtained. Addition of HNO_3 to the cold reaction mixture yields a golden-yellow crystalline product, *trans-*$[RhCl_2(en)_2]NO_3$. Evaporation of the resulting solution causes the precipitation of the more soluble, bright-yellow *cis-*$[RhCl_2(en)_2]NO_3$, equation (44). This illustrates a reaction which produces a mixture of isomers that can be readily separated by taking advantage of difference in solubilities.

$$RhCl_3 \cdot 3H_2O + 2en \cdot 2HCl \xrightarrow[\substack{H_2O \\ 100°C}]{KOH} \text{yellow solution} \quad (44)$$
$$\underset{\text{red}}{} \qquad\qquad\qquad \underset{\text{cool } HNO_3}{}$$

$$\downarrow$$

$$\underset{\text{yellow}}{cis\text{-}[RhCl_2(en)_2]NO_3} \xleftarrow[25°C]{\text{evaporate}} \underset{\text{solution}}{\text{yellow}} + \underset{\text{yellow}}{trans\text{-}[RhCl_2(en)_2]NO_3}$$

produces a mixture of isomers that can be readily separated by taking advantage of difference in solubilities.

Since in many, if not in most, attempts to produce isomeric materials one obtains a mixture of isomers or a single isomer of unknown stereochemistry, methods for determining structure are required. Several chemical tests for geometric structure are available. An elegant and absolute chemical test is the resolution of compounds of the type *cis-*$[MX_2(AA)_2]^{n+}$ into optical isomers (page 57). Another chemical test relies on the fact that typical bidentate ligands are able to span *cis* but not *trans* positions in a complex. The reaction of oxalate ion $(C_2O_4^{2-})$ with the *cis* and *trans* isomers of $[PtCl_2(NH_3)_2]$ illustrates this method. Figure 4.2 shows that the *trans* isomer forms a complex containing two oxalate ions each of which functions as a monodentate, whereas the *cis* isomer forms a complex that contains one bidentate oxalate. This approach, which has been most successful for platinum(II) complexes, is not reliable for metal complexes that readily rearrange during ligand substitution reactions.

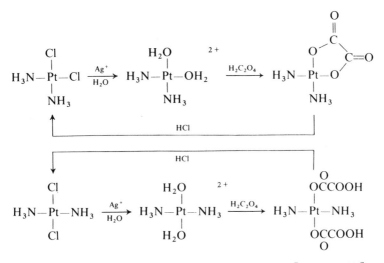

Figure 4.2 The reaction of oxalic acid with *cis*- and *trans*-[Pt(NH$_3$)$_2$Cl$_2$].

Structural assignments are now usually made on the basis of physicochemical techniques such as X-ray diffraction and spectroscopy. Another relatively simple technique involves measurements of dipole moments. The dipole moments of *cis* and *trans* isomers are often markedly different. This seems to be particularly true for certain square planar complexes, structure (45).

$$
\begin{array}{cc}
\underset{\displaystyle \overset{|}{\text{C}_6\text{H}_5}}{\overset{\displaystyle \overset{\text{P}(\text{C}_2\text{H}_5)_3}{|}}{(\text{C}_2\text{H}_5)\text{P}-\text{Pt}-\text{C}_6\text{H}_5}} & \underset{\displaystyle \overset{|}{\text{C}_6\text{H}_5}}{\overset{\displaystyle \overset{\text{C}_6\text{H}_5}{|}}{(\text{C}_2\text{H}_5)_3\text{P}-\text{Pt}-\text{P}(\text{C}_2\text{H}_5)_3}}
\end{array}
\tag{45}
$$

$$\mu = 7.2 \qquad\qquad \mu \sim 0.0$$

(in dipole units of Debye)

4.11 Preparation of optically-active compounds

Many optically-active organic molecules are present in plants and animals, and they can often be isolated and obtained in a pure form. In recent years, considerable success has been achieved in the selective synthesis of individual isomers. However, laboratory preparations of compounds that can exhibit optical activity normally yield 50–50 (racemic) mixtures of the two optical isomers and hence produce an optically-inactive material (Section 3.4). Therefore, the basic step in the laboratory preparation of an optically-active coordination compound is separation from its optical isomer. For example, racemic [Co(en)$_3$]$^{3+}$ is readily prepared by the air oxidation of a cobalt(II) salt in a medium containing excess ethenediamine and a catalytic amount of activated charcoal. Since optical isomers are very much alike, special separation techniques are required.

The most common separation techniques involve the principle that each of a pair of optical isomers will interact differently with a third optically-active material. There is a subtle structural difference between optical isomers, and this will lead one isomer to be more strongly attracted to a third dissymmetric molecule. For example, the salt $(+)$-$[Co(en)_3]((+)$-tartrate$)Cl \cdot 5H_2O$ is less soluble than $(-)$-$[Co(en)_3]((+)$-tartrate$)Cl \cdot 5H_2O$. This indicates that $(+)$-$[Co(en)_3]^{3+}$ forms a more stable crystalline lattice with a $(+)$-tartrate than does $(-)$-$[Co(en)_3]^{3+}$. Therefore, the addition of a solution containing $(+)$-tartrate anion to a concentrated solution of racemic $[Co(en)_3]^{3+}$ causes the precipitation of $(+)$-$[Co(en)_3]((+)$-tartrate$)Cl \cdot 5H_2O$. The $(-)$-$[Co(en)_3]^{3+}$ remains in solution and can be collected by the addition of I^-, which forms $(-)$-$[Co(en)_3]I_3$. This product will be contaminated with any $(+)$-$[Co(en)_3]^{3+}$ that was not removed in the tartrate precipitation.

The preferential precipitation of one of a pair of optical isomers by another optically-active compound is the principal method for resolving pairs of optical isomers. The method can be used, however, only if the isomers to be separated can be obtained as charged ions, since their precipitation as salts is required.

It is difficult to resolve nonionic compounds, but compounds enriched in one optical isomer with respect to its enantiomer have been separated by a number of methods. All of these methods involve placing a material, such as an optically-active sugar or optically-active quartz, in a column (for example, a buret) and then passing a solution of the compound to be resolved (or a gaseous compound directly) through the column. One isomer clings to the optically-active packing in the column more tightly than the other. The less tightly bound isomer moves through the column more rapidly and hence comes off the column first. The complex $[Cr(acac)_3]$ has been partially resolved by passing a solution of it through a column packed with $(+)$-lactose, a naturally occurring sugar.

Problems

1. Write appropriate balanced equations and give approximate experimental conditions for the preparation of each of the following:

$[RhONO(NH_3)_5]^{2+}$ starting with $[RhH_2O(NH_3)_5]^{3+}$
$[CoH_2O(NH_3)_5](NO_3)_3$ starting with $[CoCl(NH_3)_5]Cl_2$
$[Co(en)_3]Cl_3$ starting with $[CoCl_2(NH_3)_4]Cl$
cis-$[Pt(SCN)_2(NH_3)_2]$ starting with $K_2[Pt(SCN)_4]$
$K_2[Cu(C_2O_4)_2]$ starting with $CuCO_3$
$[Ni(DMG)_2]$ starting with $Ni(C_2H_3O_2)_2 \cdot 6H_2O$
trans-$[PtBrCl(NH_3)_4]SO_4$ starting with $[Pt(NH_3)_4]SO_4$

2 .Complete and balance the following equations. Unless otherwise stated, the reactions proceed at room temperature in water solution.

$$AgCl + S_2O_3^{2-} \rightarrow$$
$$[Ni(NH_3)_6]^{2+} + CN^- \rightarrow$$
$$[Co(H_2O)_6]^{2+} + NH_3 + NH_4Cl + SeO_2 \rightarrow$$

$$[Cr(NH_3)_5H_2O]^{3+} + OH^- \xrightarrow[\text{cold water}]{\text{fast}}$$

$$[RhCl(NH_3)_5]^{2+} + Hg^{2+} + H_2O \xrightarrow{\text{heat}}$$

3. Identify the colored species designated by capital letters. Excess ammonia was added to a pink aqueous solution (A) in the absence of air to give a pale-blue solution (B). In the presence of air, the color of this solution slowly changed and eventually became rose (C). The addition of charcoal to the boiling solution (C) resulted in a yellow solution (D). Addition of excess HCl to this yellow solution caused the precipitation of a yellow-orange crystalline product (E).

4. (*a*) By making use of the *trans*-effect phenomenon in platinum(II) complexes, show how the following complexes may be prepared starting with

$$\begin{array}{ccc}
\overset{\displaystyle NH_3}{\underset{\displaystyle NH_3}{Cl-Pt-py}}{}^{+} &
\overset{\displaystyle NH_3}{\underset{\displaystyle py}{Cl-Pt-NH_3}}{}^{+} &
\overset{\displaystyle C_2H_4}{\underset{\displaystyle NH_3}{Cl-Pt-Cl}}
\end{array}$$

K_2PtCl_4. (*b*) Show the method of preparation for all of the possible isomers of $[Pt(I)(Cl)(py)(NH_3)]$.

References

The preparations of inorganic compounds have been published in volumes of *Inorganic Syntheses* starting in 1939. The 22 volumes now available include many examples of metal complexes. Detailed specific directions are given.

Inorganic Syntheses (McGraw-Hill, New York, 22 vols, 1939–1983).

Schlessinger, G. G. *Inorganic Laboratory Preparations* (Chemical Publishing, New York, 1962).

Jolly, W. L. (ed.) *Preparative Inorganic Reactions* (Interscience, New York, 7 vols, 1964–1971; "The Application of Reaction Mechanisms to the Syntheses of Coordination Compounds", Burmeister, J. L. and Basolo, F. Vol. 5, p. 1 (1968)

Angelici, R. J. *Synthesis and Technique in Inorganic Chemistry* (Saunders, Philadelphia, 1977).

Chapter 5

Complex Ion Stability

To understand the solution chemistry of metals, we must know the nature and stability of the complexes that metal ions can form with the solvent and with potential ligands in solution. Research provides an insight into factors that contribute to stability of metal complexes. Many important applications of this information have been made. Recall that in qualitative analysis certain precipitates are dissolved by using an appropriate complexing agent. The use of hypo, $Na_2S_2O_3$, in the fixing process in the photographic industry is effective because insoluble silver halide in the film emulsion is dissolved owing to the formation of the stable and soluble silver thiosulfate complex $[Ag(S_2O_3)_2]^{3-}$. Addition of complexing agents to hard water results in the generation of stable and soluble complexes of objectionable metal ions, such as calcium, which in turn prevents precipitation of insoluble metal salts of common soaps (soap scum).

When ammonia is added to a solution of a copper(II) salt, there is a rapid reaction in which water coordinated to Cu^{2+} is replaced by ammonia. Although the product of this reaction is normally represented as $[Cu(NH_3)_4]^{2+}$, in fact, a variety of products result, the relative amount of each species depending upon the concentrations of copper(II) ion and ammonia reactions (1) to (4). Figure 5.1 is a plot of the percentage of each copper(II) ammonia species in solution against the concentration of free ammonia. This plot indicates that $[Cu(NH_3)_4(H_2O)_2]^{2+}$ is the predominant complete ion in solutions containing 0.01 to 5 M free ammonia. Outside this range, however, the other ammine complexes are more abundant.

$$[Cu(H_2O)_6]^{2+} + NH_3 \rightleftharpoons [CuNH_3(H_2O)_5]^{2+} + H_2O \tag{1}$$

$$[CuNH_3(H_2O)_5]^{2+} + NH_3 \rightleftharpoons [Cu(NH_3)_2(H_2O)_4]^{2+} + H_2O \tag{2}$$

$$[Cu(NH_3)_4(H_2O)_2]^{2+} + NH_3 \rightleftharpoons [Cu(NH_3)_3(H_2O)_3]^{2+} + H_2O \tag{3}$$

$$[Cu(NH_3)_3(H_2O)_3]^{2+} + NH_3 \rightleftharpoons [Cu(NH_3)_4(H_2O)_2]^{2+} + H_2O \tag{4}$$

On a strictly statistical basis, one would expect that the relative number of H_2O and NH_3 molecules surrounding Cu^{2+} would be the same as the relative

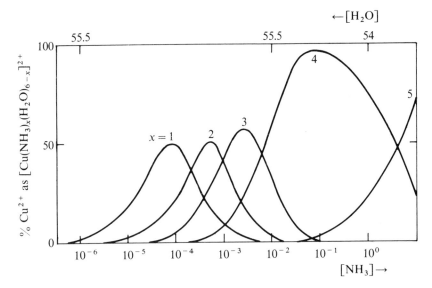

Figure 5.1 A plot of the percentage of Cu^{2+} present in the form of various ammine complexes for solutions containing different concentrations of free ammonia at 18°C. (For example, at a free NH_3 concentration of 1.0×10^{-2} M, the Cu^{2+} is present as 62 per cent $[Cu(NH_3)_4(H_2O)_2]^{2+}$, 34 per cent $[Cu(NH_3)_3(H_2O)_3]^{2+}$, and 4 per cent $[Cu(NH_3)_2(H_2O)_4]^{2+}$).

number of NH_3 and H_2O molecules in solution; that is, if the solution contains equal numbers of NH_3 and H_2O molecules, the predominant copper(III) species should be $[Cu(NH_3)_3(H_2O)_3]^{2+}$. This type of statistical ligand distribution is not observed. Metal ions exhibit marked preferences for certain ligands; for example, copper(II) ion coordinates NH_3 in preference to H_2O. Nonetheless, statistical considerations are also important in that an increase in NH_3 concentration produces copper(II) species containing a larger number of coordinated NH_3 molecules.

In some cases, the preferences of metals for certain ligands is easily understood; it seems reasonable that positive metal ions should prefer anionic ligands to neutral or to positively charged ones. In general, however, the factors that determine which ligand will coordinate best with a given metal ion are numerous, complicated and not completely understood. Some of these factors will be discussed later in this chapter.

The properties of a metal ion in solution are dependent on the nature and number of the groups (ligands) surrounding the metal. The number and type of such groups cannot be predicted easily. Consequently, many studies have been made to establish the composition of the coordination spheres of metal ions in solution in the presence of a wide variety of possible ligands. In discussions of the solution behavior of coordination compounds, it is normally assumed that the solvent is water, but nonaqueous solvents will dissolve certain coordination compounds and are being more widely used. In these solvents, solute species are surrounded by solvent molecules, and complexation reactions then involve the replacement of solvent molecules by other ligands. In principle, equilibrium behavior in nonaqueous solvents can be handled in a manner analogous to that of water. The limited solubility of ionic species in most nonaqueous solvents, problems associated with the lack of dissociation of salts (ion pairing) in most nonaqueous solvents, and the convenience of aqueous systems have resulted in the use of aqueous media in most equilibrium studies. In the following discussion, aqueous equilibria will be specifically considered but, with certain modifications, the same treatment can be applied in other solvents.

5.1   Stability constants

It has been found that in a reaction mixture at equilibrium at a certain temperature the product of the activities of the products divided by the

$$aA + bB + \cdots \rightleftharpoons cC + dD + \cdots$$

product of the activities of the reactants is equal to a constant called the *equilibrium constant* for the given reaction, equation (5).

$$K = \frac{a_C{}^c a_D{}^d \cdots}{a_A{}^a a_B{}^b \cdots} = \text{constant} \qquad (5)$$

The *activity* of a species A is the product of its concentration and an *activity coefficient* γ_A.

$$a_A = [A]\gamma_A$$

The activity coefficient has a value of unity in very dilute solutions, so that under such conditions, concentrations and activities are numerically equal. In the 0.01 to 5 M solutions most commonly used in the laboratory, activity coefficients are less than 1, and hence activities are lower than concentrations.

That the activity of a species in solution is less than its concentration is interpreted as indicating that the species cannot act independently but is under the influence of other solute particles: hence its effective concentration is decreased. In subsequent discussions of equilibrium constants, we shall frequently replace activities with concentrations. Note that this involves the assumption of unit activity coefficients, and thus will be quantitatively accurate only in very dilute solution.

Metal complexes are formed in solution by stepwise reaction, and equilibrium constants can be written for each step, equations (6), (7).

$$Ag^+ + NH_3 \rightleftharpoons Ag(NH_3)^+ \qquad K_1 \qquad (6)$$

$$Ag(NH_3)^+ + NH_3 \rightleftharpoons [Ag(NH_3)_2]^+ \qquad K_2 \qquad (7)$$

The water molecules that make up the hydration sphere of an aqueous metal ion are often omitted in writing the reaction equation. Moreover, the solvent water molecules involved in the reaction are never included in the equilibrium expression.

The equilibrium constants K_1 and K_2, equation (8), are called *step-wise*

$$K_1 = \frac{[Ag(NH_3)^+]}{[Ag^+][NH_3]} \qquad K_2 = \frac{[Ag(NH_3)_2{}^+]}{[Ag(NH_3)^+][NH_3]} \qquad (8)$$

stability constants. The larger the value of the constant, the greater the concentration of the complex species at equilibrium.

A second type of equilibrium constant β, called an *over-all stability constant*,

$$\beta_1 = \frac{[Ag(NH_3)^+]}{[Ag^+][NH_3]} \qquad \beta_2 = \frac{[Ag(NH_3)_2{}^+]}{[Ag^+][NH_3]^2} \qquad (9)$$

is also used, equation (9). Since the K's and β's describe exactly the same chemical systems, they must be related to each other, equation (10).

$$\beta_2 = \frac{[Ag(NH_3)_2{}^+]}{[Ag^+][NH_3]^2} = \frac{[Ag(NH_3)_2{}^+]}{[Ag^+][NH_3][NH_3]} \cdot \frac{[Ag(NH_3)^+]}{[Ag(NH_3)^+]}$$

$$= \frac{[Ag(NH_3)_2{}^+]}{[Ag(NH_3)^+][NH_3]} \cdot \frac{[Ag(NH_3)^+]}{[Ag^+][NH_3]} = K_2 K_1 \qquad (10)$$

From equation (9) it is seen that $\beta_1 = K_1$, and from equation (10) that $\beta_2 = K_1 K_2$. The general relationship is given by equation (11).

$$\beta_n = K_1 K_2 \ldots K_n. \qquad (11)$$

The numerical value of a stability constant describes the relative concentrations of species at equilibrium. Large stability constants indicate that the concentration of a complex is much greater than the concentrations of the components of which it is made. A complex is said to be stable if the equilibrium constant describing its formation is large. We shall see in Chapter 6 that this does not necessarily imply that the compound is slow to react or that the ligands are resistant to replacement by ligands other than water.

5.2   Factors that influence complex stability

The term "stable complex" is defined in terms of the equilibrium constant for the formation of the complex. The *equilibrium constant* of a reaction is a measure of heat released in the reaction and the entropy change during the reaction. The greater the amount of heat evolved in a reaction the more stable are the reaction products. The entropy of a system is a measure of the amount of disorder. The greater the amount of disorder in the products of a reaction relative to the reactants, the greater the increase in entropy during the reaction and the greater the stability of the products. Now let us consider separately the influence of the heats and entropies of complexation reactions on the stability of the complex formed.

Relative stabilities of many complexes can be understood in terms of a simple electrostatic model. The predictions of this model are related primarily to the heat evolved in the formation of the complexes. We are familiar with the observation that oppositely charged particles attract one another, whereas like charged particles repel one another. Moreover, the repulsion or attraction depends upon the distance between the centers of the particles, being greater as the particles approach one another.

One would, therefore, predict that the most stable complexes would be made up of oppositely charged ions and, moreover, that the greater the charge on the ions and the smalller the ions the greater should be the stability. Small ions are favored because their centers can be closer together. From this point of view the stability of complexes should increase with the charge on metal ions. One illustration of this behavior is the increase of stability of hydroxide complexes with an increase in charge of the metal ion.[1] The stability of

$$K_{LiOH} = 2 \qquad K_{MgOH^+} = 10^2 \qquad K_{YOH^{2+}} = 10^7 \qquad K_{ThOH^{3+}} = 10^{10}$$

complexes of metal ions having the same charge should increase as the ionic radius decreases. The stability constants of MOH^+ for the alkaline-earth metals illustrate this trend. From these data one can see that the stability

$$K_{BeOH^+} = 10^7 \qquad K_{MgOH^+} = 120 \qquad K_{CaOH^+} = 30 \qquad K_{BaOH^+} = 4$$

constants of YOH^{2+} and $BeOH^+$ are about the same. Thus a very small doubly charged cation can form complexes with stability comparable to that of complexes of larger, more highly charged cations.

The stabilities of high-spin complexes of the M^{2+} ions between Mn^{2+} and Zn^{2+} with a given ligand frequently vary in the order $Mn^{2+} < Fe^{2+} < Co^{2+} < Ni^{2+} < Cu^{2+} > Zn^{2+}$. This order, sometimes called the *natural order of stability*, is relatively consistent with the charge-to-radius concept, since the radii of the ions vary in the order $Mn^{2+} > Fe^{2+} > Co^{2+} > Ni^{2+} < Cu^{2+} < Zn^{2+}$. Both the variation in size of the cation and the order of stability can be explained in terms of the crystal field stabilization energy (CFSE) for these complexes (Section 2.5).

Figure 5.2 indicates the relative stability of high-spin octahedral $[M(II)L_6]$ complexes of first-row transition elements as predicted by CFT. The d^3 and d^8 systems will be the most stable with respect to their neighbors, since they have the greatest CFSE. As one progresses from Ca^{2+} to Zn^{2+} complexes, a general increase in stability is observed. The order of stability predicted by CFT, and presented in Figure 5.2, parallels the natural order of stability for complexes of

---

[1] Stability constants reported in this book are values at 25°C and are activity constants unless otherwise specified. Since literature values are often not in agreement, only one significant figure will generally be used.

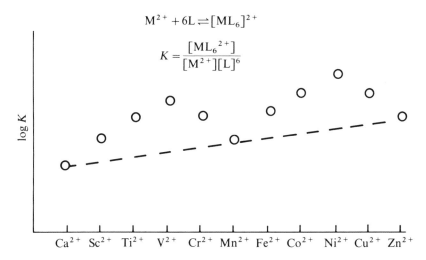

Figure 5.2 The logarithms of stability constants for a series of $[ML_6]^{2+}$ complexes as predicted by crystal field theory. Without the concepts of crystal field theory the values would be expected to increase fairly regularly along the dashed line. Experimental results show the double maxima represented by the circles.

these metals except for Cu^{2+}. This discrepancy is not completely understood, but it is certainly related to the fact that Cu^{2+} complexes assume a distorted octahedral structure.

A number of important ligands are neutral molecules (H_2O, NH_3, H_2S, etc.). In terms of electrostatics, these ligands are bound to metal ions through the attraction between the negative end of the ligand dipole and the metal cation, structure (I). The more polar the ligand, the greater should be the force binding the ligand and metal ion. Water is the most polar of the common ligands and hence would be expected to form metal complexes of greater stability than other neutral ligands. The fact that water is the best solvent for many salts is, in part, a result of the stable complexes that it forms with metal ions.

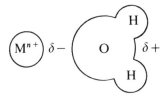

It has been observed that the greater the base strength of a ligand, the greater is the tendency of the ligand to form stable metal complexes. The base strength of a molecule is a measure of the stability of the "complex" that the molecule forms with H^+. It is reasonable that ligands which bind H^+ firmly should also form stable complexes with metal ions. From this point of view, F^-

should form more stable complexes than Cl^-, Br^-, or I^-, and NH_3 should be a better ligand than H_2O, which, in turn, should be better than HF. This predicted behavior is observed for the alkali and alkaline-earth metals and for other electropositive metals such as the first-row transition elements and the lanthanide and actinide elements. These metals are often called *class a metals*.

The simple electrostatic approach that has been outlined is successful in explaining the observed stability of many metal complexes and also in predicting the stabilities of other compounds. It is particularly effective for complexes of class *a* metal ions. In complexes of *class b metal ions*—ions of the more electronegative metals such as Pt, Au, Hg, and Pb and some lighter transition metals in low oxidation states—the electrostatic contributions are still important but other factors also play a role. In particular, crystal field effects and covalent bonding are also important. For example, we find that certain transition metal ions form very stable complexes with ligands such as CO, CN^-, C_2H_4, and $P(CH_3)_3$, whereas these are very poor ligands for nontransition metals.

Metal–ligand covalent bonding becomes increasingly important in complexes of relatively electronegative metals such as those in the copper and zinc families and tin and lead. For these metals, an electrostatic approach to stability is often ineffective: for example, silver forms insoluble halide salts, AgX, and stable halide complexes, AgX_2^- and AgX_3^-, in which the order of stability is $I^- > Br^- > Cl^- \gg F^-$. The stability constants for reaction (12) are $K_{AgF} = 2$, $K_{AgCl} = 2 \times 10^3$, $K_{AgBr} = 5 \times 10^4$, $K_{AgI} = 4 \times 10^6$.

$$Ag^+ + X^- \rightleftharpoons AgX \text{ (aqueous)} \qquad (12)$$

This behavior is attributed to covalent character in the Ag—X bond, which increases as one goes from F^- to I^-.

Mercury, lead, bismuth and other transition and post-transition metals form water-insoluble sulfide salts. The precipitation of these sulfides is part of the traditional method for the qualitative determination of metals. The formation of a precipitate can be considered as the formation of a zero-charged, water-insoluble complex; the formation of the sulfide precipitates indicates that these metals prefer sulfur-containing ligands to oxygen-containing ligands (in this case S^{2-} to O^{2-}). This preference for sulfur can be attributed to the presence of considerable covalent character in the metal–sulfur bonds. This phenomenon was recognized more than a century ago by Professor Berzelius (1779–1848) of Upsala University who pointed out that certain metals were found in the Earth's crust as oxide minerals and other metals occur as sulfide minerals.

The class *b* metals are characterized by the presence of a number of *d* electrons beyond a noble-gas core. These *d* electrons can be used to π bond with ligand atoms, and the presence of such π bonding gives rise to many of the properties of the class *b* metals. The most stable complexes of these metals are formed with ligands that can accept electrons from the metal, that is, ligands with vacant d orbitals such as $P(CH_3)_3$, S^{2-}, and I^-, or ligands with molecular orbitals into which electrons can be delocalized, such as CO and CN^- (Figure 2.21). Class a and b elements, therefore, form stable compounds with different types of ligands. Class a elements prefer O- and N-containing ligands and F^-. Class b elements form more stable complexes with the heavier elements of the N, O, and F families.

Class a metals are called *hard acids* and class b metals, *soft acids*. Soft acids and bases are generally highly polarizable whereas hard acids and bases are not. Ligand atoms such as N, O, and F are *hard bases* and those similar to P, S, and I are *soft bases*. The most stable complexes generally result from hard acid–hard base and soft acid–soft base combinations.

A completely adequate explanation of the stability of metal complexes is difficult, since little heat is evolved when water molecules are displaced by other ligands, reaction (13).[1] A variety of relatively small effects such as π

$$[M(H_2O)_x]^{n+} + yL \rightarrow [M(H_2O)_{x-y}L_y]^{n+} + yH_2O \qquad (13)$$

bonding, crystal field stabilization, and increased covalent character in the metal–ligand bond can provide enough energy to alter the behavior that one might consider "normal".

Entropy changes also play an important role in determining complex stability. Reactions in which positive ions and negative ligands interact to form complexes of lower charge, reaction (14) proceed with a large increase

$$[M(H_2O)_n]^{3+} + L^- \rightarrow [M(H_2O)_{n-1}L]^{2+} + H_2O \qquad (14)$$

in entropy; this is a major factor in the stability of the resulting complex. The large entropy change arises mainly because each charged reactant has an ordered solvation sphere. The resulting products, having a lower charge, will produce considerable less ordering of the solvent. Fortunately, the factors that produce this increase in entropy are the same as those which produce stability in terms of electrostatics. Hence, the electrostatic prediction of high stability due to the interaction of small highly charged ions may be accurate largely because of the entropy effect.

Entropy considerations are very important in two other ways. In the formation of a metal complex $[ML_6]^{n+}$ from $[M(H_2O)_6]^{n+}$, each replacement of an additional H_2O molecule by another ligand L becomes more difficult. For example, successive stepwise stability constants for the reaction (15) are $K_1 = 5 \times 10^2$, $K_2 = 1.3 \times 10^2$, $K_3 = 4 \times 10^1$, $K_4 = 1.2 \times 10^1$, $K_5 = 4$, and

$$[Ni(H_2O)_6]^{2+} + 6NH_3 \rightleftharpoons [Ni(NH_3)_6]^{2+} + 6H_2O \qquad (15)$$

$K_6 = 0.8$. This effect arises, at least in part, from the statistics of substitution processes (an entropy consideration). The replacement of one water molecule by ammonia removes one possible site for the coordination of additional

---

[1] The heat evolved when a gaseous metal ion combines with ligands is quite large.

ammonia molecules. Moreover, the larger the number of ammonia molecules present in the complex, the greater the probability of their replacement by water. Both of these factors reduce the stability of the more highly substituted complexes. Other factors, such as steric repulsion between bulky ligands and electrostatic repulsion as anionic ligands replace water molecules on a positive metal ion, can also retard the coordination of additional ligands.

There are, however, a few examples in which the initial complexes are less stable than their more highly substituted relatives. Deviations from a regular decreasing trend of stepwise stability constants have been interpreted in some cases as indicating a change in the coordination number of the metal ion. The stability constants for $[CdBr_4]^{2-}$ are $K_1 = 2 \times 10^2$, $K_2 = 6$, $K_3 = 0.6$, $K_4 = 1.2$. Because the coordination numbers of cadmium in the hydrated ion and in $[CdBr_4]^{2-}$ are probably 6 and 4, respectively, the large value of K_4 may indicate that reaction (16), corresponding to this constant, involves a change

$$[Cd(H_2O)_3Br_3]^- + Br^- \rightleftharpoons [CdBr_4]^{2-} + 3H_2O \qquad (16)$$

in coordination number as well as the addition of Br^-.

The second very important entropy-induced effect is the great stability of metal chelates (see definition of chelate in Section 1.3). Both ammonia and ethylenediamine (en) coordinate with metals through amine nitrogens; in terms of the heat evolved in the complexation reaction two molecules of NH_3 have been shown to be about equivalent to one en molecule. However, en complexes are considerably more stable than their NH_3 counterparts for example, $[Ni(NH_3)_6]^{2+}$, $K_1K_2 = 6 \times 10^4$; $K_3K_4 = 5 \times 10^2$; $K_5K_6 = 3$. $[Ni(en)_3]^{2+}$, $K_1 = 5 \times 10^7$; $K_2 = 2.2 \times 10^6$; $K_3 = 3.6 \times 10^4$). It has been experimentally demonstrated that the unusual stability of the en compounds is due to a more favorable entropy associated with their formation.

Chelating ligands, in general, form more stable complexes than their monodentate analogs. This *chelate effect* can be explained in terms of the favorable entropy for the chelation process. One can understand the favorable entropy in a qualitative fashion. The probability of replacement of a coordinated water molecules by either an NH_3 or an en molecule should be about equal. The replacement of a second water molecule by the other amine group in the coordinated en, however, is much more probable that its replacement by a free NH_3 molecule from solution, since the en is already tied to the metal ion and the free end of the molecule is in the immediate vicinity of the H_2O that it will replace. Thus, the formation of $[Ni(en)(H_2O)_4]^{2+}$ is more probable than the formation of the less stable $[Ni(NH_3)_2(H_2O)_4]^{2+}$.

Another way of visualizing the more favorable entropy change is to realize that a process in which the number of independent particles increases proceeds with an increase in entropy (the larger the number of particles, the greater is the possible disorder). In the coordination of one en molecule, two H_2O molecules are freed; this process should proceed with a favorable entropy.

Tri-, quadri-, and other polydentate ligands can replace three, four, or more

Table 5.1 The effect of chelation on the stability of complexes[a]

Complex	β's					
	β_1	β_2	β_3	β_4	β_5	β_6
$[Ni(NH_3)_6]^{2+}$	5×10^2	6×10^4	3×10^6	3×10^7	1.3×10^8	1.0×10^8
	β_1			β_2		β_3
$[Ni(en)_3]^{2+}$	5×10^7			1.1×10^{14}		1.4×10^{18}
		β_1			β_2	
$[Ni(dien)_2]^{2+b}$		6×10^{10}			8×10^{18}	

[a] All β's were measured in 1 M KCl at 30°C. β's for complexes containing the same amount of coordinated nitrogen are in the same column. To see the effect of chelation the β's in the same column should be compared.
[b] dien is $H_2NCH_2CH_2NHCH_2CH_2NH_2$.

coordinated water molecules, respectively, to form even more stable complexes. The stability constants for some complexes of Ni^{2+} with polydentate ligands are presented in Table 5.1. Ethylenediaminetetraacetate (EDTA), a sexidentate ligand (see structure XXVII on page 58), forms stable complexes with a wide variety of metal ions including the alkaline-earth metals (which form very unstable complexes with unidentate ligands). This compound is used commercially as a *sequestrant*, a reagent that will form complexes with metal ions and in this way control their concentration in solution. For example, EDTA will complex calcium ion very efficiently and is, therefore, an excellent water-softening agent; it is also used as an analytical reagent.

In basic solution, EDTA reacts quantitatively with many metal ions to produce the metal complex, reaction (17). A solution of EDTA can, therefore,

$$EDTA^{4-} + [M(H_2O)_6]^{2+} \rightarrow [MEDTA]^{2-} + 6H_2O \qquad (17)$$

be used as a titrant for the volumetric determination of many metals. Several excellent indicators have been developed to detect the end points; such volumetric techniques are now common in quantitative chemical analysis.

Metal chelates contain rings of atoms, structure II. The stability of the

II

complex ion has been observed to be dependent on the number of atoms in the ring. In general, it has been observed that for ligands that will not contain double bonds in the ring formed, those that form five-membered metal–chelate rings give the most stable products. Ligands that contain double bonds, such as acetylacetone, form very stable metal complexes containing six-membered rings. Chelate rings that contain either four atoms or more than six atoms are known to be less stable, except for bridged complexes, structure III.

$$\begin{array}{c} \diagdown | \quad X \quad | \diagup \\ M \quad\quad M \\ \diagup | \quad X \quad | \diagdown \end{array} \qquad X = OH^-,\ Cl^-,\ \text{etc.}$$

III

Chelate ligands which completely surround a metal ion are called *cryptands* and form very stable complexes called *cryptates*. For example, the crown ether, 15-crown-5, forms a very stable complex with Na^+, structure IV.

IV

Ions such as Na^+ and K^+ form such unstable complexes with conventional ligands that the stability of their cryptate complexes is particularly noteworthy. This stability is quite sensitive to a match between the sizes of the cation and the hole in the interior of the ligand. For cyclic ethers with the general formula $(-O-CH_2-CH_2-)_n$, the ether with $n = 4$ forms the most stable Li^+ complex; the one with $n = 5$ forms the most stable Na^+ complex; and the one with $n = 6$ forms the most stable K^+ complex. Placing an ion inside an organic ring gives it much greater solubility in nonpolar solvents. It is thought that the transport of Na^+ and K^+ ions through biological membranes (which are nonpolar in nature) involves the formation of complexes of this general type. The use of cryptands to solubilize metal salts in nonpolar solvents has led to much interesting chemistry. For example, a salt containing the anion Na^- and a Na^+ cryptate cation can be isolated from a solution of sodium metal in ethylamine containing a suitable cryptand.

There is a lot of information available on the stabilities of metal complexes. This permits an assessment of the various factors that influence the stability of a metal complex. Some of these factors have already been discussed, but it may be helpful to summarize them briefly. First, the stability of a complex obviously depends on the nature of the metal and of the ligand. With reference to the metal, the following factors are important:

(1) *Size and charge*. Because of the significant influence of electrostatic forces in these systems, the smaller the size and the larger the charge of a metal ion,

the more stable are the metal complexes. Thus stability is favored by a large charge-to-radius ratio of the metal ion.

(2) *Crystal field effects.* The crystal field stabilization energy (CFSE) plays an important role in the stability of transition–metal complexes and seems to be responsible for the order of stability of the complexes of the first-row transition metals (Figure 5.2).

(3) *Class a and class b metals.* The more electropositive metals, for example Na, Ca, Al, the lanthanides, Ti, and Fe, belong to class *a*. The less electropositive metals, for example Pt, Pd, Hg, Pb, and Rh, belong to class *b*. Class *a* metals form their most stable complexes with ligands in which the donor atom is N, O, or F; class *b* metals prefer ligands in which the donor atom is one of the heavier elements in the N, O, or F families. The stability of the complexes of the class *b* metals is believed to result from an important covalent contribution to metal–ligand bonds and from the transfer of electron density from the metal to the ligand via π bonding.

The following factors are important when considering the role of the ligands in determining the stability of metal complexes:

(1) *Base strength.* The greater the base strength of a ligand toward H^+, the greater is the tendency of the ligand to form stable complexes with class *a* metals.

(2) *Chelate effect.* The stability of a metal chelate is greater than that of an analogous nonchelated metal complex, for example, $[Ni(en)_3]^{2+} > [Ni(NH_3)_6]^{2+}$. The more extensive the chelation, the more stable the system; recall the very stable complexes of the sexidentate EDTA.

(3) *Chelate ring size.* The most stable metal chelates contain saturated ligands that form five-membered chelate rings or unsaturated ligands that form six-membered rings.

(4) *Steric strain.* Because of steric factors, large bulky ligands form less stable metal complexes than do analogous smaller ligands; for example, $H_2NCH_2CH_2NH_2$ forms more stable complexes than $(CH_3)_2NCH_2CH_2N-(CH_3)_2$. The strain is sometimes due to the geometry of the ligand coupled with the stereochemistry of the metal complex. For example, $H_2NCH_2CH_2-NHCH_2CH_2NHCH_2CH_2NH_2$ can coordinate its four nitrogens at the corners of a square, but $N(CH_2CH_2NH_2)_3$ cannot; thus the straight-chain tetramine forms more stable complexes with Cu^{2+} than does the branched-chain amine, which is unable to assume the preferred square planar geometry, structure V.

V

5.3   Stabilization of unusual oxidation states by coordination

A characteristic of transition metals is their ability to exhibit a variety of oxidation states. By surrounding metal atoms with appropriate ligands it has been possible to produce compounds in which most transition elements exhibit all oxidation states from some maximum (which is related to its position in the Periodic Table) to -1 or -2. For example, compounds in which iron exhibits all oxidation states from $+6$ (K_2FeO_4) to -2 ($Na_2[Fe(CO)_4]$) are known. The ability to prepare compounds in which elements exhibit unusual oxidation states is a result of our present knowledge of factors which stabilize these states. (Note that oxidation states are assigned by a simple arbitrary procedure for counting electrons. The oxidation state of an atom has some relation to the charge on an atom but is not an accurate measure of it. Even with sophisticated modern instruments actual charges on atoms are difficult to measure.)

Transition metals, in general, tend to give up electrons to other elements when they form compounds. As a consequence, they normally have a positive charge in compounds. The magnitude of this charge depends on the nature of the metal and the atoms to which it is bonded, but the actual charge on the metal probably does not vary over a great range: it certainly varies less than the assigned oxidation states. It is, therefore, reasonable to assume that ligands which stabilize high positive oxidation states are able to do so, at least in part, because they release electrons back to the metal (through π-bonding) in order to reduce the positive charge on the metal. Oxide and fluoride ions are the most commonly used to stabilize high oxidation states and both have pairs of electrons in p orbitals suitable for π-bonding with vacant orbitals on a metal.

Compounds in which metals have uncommonly high oxidation states include OsO_4, $K_2[NiF_6]$, and $KCuO_2$. The stability of these compounds is also related to the fact that fluoride and oxide ions are difficult to oxidize and can survive in the presence of strong oxidants such as osmium(VIII), nickel(IV), and copper(III). Some rather oxidizable ligands such as $S_2C_2(CF_3)_2{}^{2-}$ (see structure II in Chapter 3) also form complexes in which metals exhibit apparent high oxidation states, for example $[Ni(S_2C_2(CF_3)_2)_2]$. It has been suggested that in this and related cases, oxidation occurs primarily on the ligand (which can give up electrons), and the metal remains in a common oxidation state, in this case nickel(II).

Low oxidation states are stabilized by ligands which can accept electrons from the metal (through π-bonding) and in this way increase the positive charge on the metal. These ligands are called π acids. The ligands NO^+ and CO are very effective in this regard; CN^-, bipy, phosphines, and arsines are also effective. Examples of compounds in which the metal has an unusually low oxidation state are $Na[V(CO)_6]$, $[Cr(bipy)_3]$, and $K_4[Ni(CN)_4]$.

The electrode potentials for metal ions change markedly as the type of ligand is changed (Table 5.2). As water is replaced by CN^-, EDTA, or NH_3 in complexes of Fe^{2+} and Co^{2+}, the tendency for oxidation to the M^{3+} state is

Table 5.2 Electrode potentials for some cobalt and iron complexes

Reaction	Electrode potential (*volts*)
$[Fe(H_2O)_6]^{3+} + e \rightarrow [Fe(H_2O)_6]^{2+}$	0.77
$[Fe(CN)_6]^{3-} + e \rightarrow [Fe(CN)_6]^{4-}$	0.37
$[FeEDTA]^- + e \rightarrow [FeEDTA]^{2-}$	−0.12
$[Co(H_2O)_6]^{3+} + e \rightarrow [Co(H_2O)_6]^{2+}$	1.84
$[Co(NH_3)_6]^{3+} + e \rightarrow [Co(NH_3)_6]^{2+}$	0.10

markedly enhanced. These ligands form very much more stable complexes with M^{3+} ions than with M^{2+} ions; this provides the driving force for the oxidation. This is particularly true in Co^{2+} systems. The complex $[Co(H_2O)_6]^{3+}$ will oxidize H_2O to O_2; in contrast, aqueous solutions of Co^{2+} salts are readily oxidized by atmospheric O_2 to $[Co(III)L_6]$ complexes in the presence of ligands such as NH_3, CN^-, or $NO_2{}^-$. The large change in electrode potential that results from the presence of these ligands is largely due to the fact that the ligands provide a greater CF than H_2O; this favors the conversion of the high-spin d^7 Co^{2+} complexes to the highly CF-stabilized low-spin Co^{3+} d^6 complexes.

5.4 Determination of stability constants

Our comments on the stability of metal complexes have evolved from the study of stability constant data. The experimental determination of stability constants is an important but often difficult task. Perhaps the greatest problem in equilibrium measurements is to determine which species are actually present in solution. Equilibrium constants have been measured by many different methods. Usually a solution of the metal ion and ligand is prepared, sufficient time is allowed for the system to come to equilibrium, and the concentrations of the species in solution are then measured. From these equilibrium concentrations one can calculate the equilibrium constant using an expression such as equation (18).

$$A + B \rightleftharpoons C$$

$$K = \frac{[C]}{[A][B]} \tag{18}$$

The measurement of equilibrium concentrations of species is complicated by the fact that the measurement must not disturb the equilibrium. For example, in reaction (19) one could not measure the Cl^- concentration in

$$[Co(H_2O)_6]^{2+} + Cl^- \rightleftharpoons [Co(H_2O)_5Cl]^+ + H_2O \tag{19}$$

solution by the amount of precipitation of AgCl. The addition of Ag^+ would not only precipitate free chloride ion but also remove the Cl^- from the cobalt complex.

Concentrations of species in solution can be measured by a variety of methods that do not disturb the equilibrium under study: the most common are probably spectroscopy and electroanalysis. Spectroscopy involves the absorption of light by the species to be studied; electroanalysis involves the electrochemistry of the system being investigated. Spectroscopic techniques can be illustrated by studies on the Fe^{3+}—NCS^- equilibrium (20). Iron(III) ion

$$Fe^{3+} + NCS^- \rightleftharpoons Fe(NCS)^{2+} \qquad (20)$$

and thiocyanate ion individually are almost colorless in solution. The $FeNCS^{2+}$ complex, however, exhibits a bright red-orange color; that is, the individual ions do not absorb visible light, whereas $FeNCS^{2+}$ does. The intensity of the red-orange color depends directly on the concentration of $FeNCS^{2+}$ and can be used to measure that concentration in solution. This measurement is actually more complicated than the description infers, since species such as $[Fe(NCS)_2]^+$ also exist. It is possible to do the experiment with an excess of Fe^{3+} ion so that the concentrations of higher complexes are negligible and essentially the only species present are Fe^{3+} and $FeNCS^{2+}$.

The simplest electroanalytical technique for the determination of stability constants is one that makes use of the glass electrode. This device is the fundamental component of the common laboratory pH meter used to determine the activity of H^+ in solution; therefore, equilibria studied with the device must involve changes in $[H^+]$. Professor J. Bjerrum, of the University of Copenhagen, when a student, determined the stability constants of a variety of NH_3 complexes using this technique.

Many other experimental methods are available. For example, methods that use radioactive isotopes, or employ liquid–liquid extraction or ion exchange are fairly common. Virtually any technique that can be used to determine concentration can be, and has been used, to determine stability constants.

The stability of metal complexes in solution is one important aspect of the solution chemistry of metals. The structure of the solvent, the formulas and nature of the hydration spheres of the solute species present, and the reactions between species and their neighbors and the equilibria resulting therefrom have interested scientists for many years. A great deal of information has been gathered, and elaborate theories have been developed to explain the experimental data. In spite of all this effort, solution chemistry remains a challenging research field. Basic information, such as reliable stability constants for many species, especially the unstable and the very stable, is not yet available.

Problems

1. (*a*) Write all the stepwise (K's) and overall (β's) stability constants for the following reactions. (*b*) Estimate which of all the stepwise constants will be the greatest. (*c*) Predict which of all the stepwise constants will be the smallest.

$$Ni^{2+} + 4CN^- \rightleftharpoons [Ni(CN)_4]^{2-} \text{ (diamagnetic)}$$
$$Ag^+ + 2NH_3 \rightleftharpoons [Ag(NH_3)_2]^+$$
$$Cr^{3+} + 3en \rightleftharpoons [Cr(en)_3]^{3+}$$
$$Fe^{3+} + 4Cl^- \rightleftharpoons [FeCl_4]^-$$

2. The following compounds were dissolved in 100 cm^3 of water: 1.00×10^{-3} moles of dien ($H_2NCH_2CH_2NHCH_2CH_2NH_2$) and 5.00×10^{-3} moles of $Ni(ClO_4)_2$. (*a*) Calculate the concentration of $[Ni(dien)]^{2+}$ in solution (you can assume that the concentration of $[Ni(dien)_2]^{2+}$ is much less than either Ni^{2+} or $[Ni(dien)]^{2+}$). (*b*) Calculate the concentration of Ni^{2+} in solution. (*c*) Calculate the concentration of $[Ni(dien)_2]^{2+}$ in solution.

$$K_1 = 5.0 \times 10^{10} \qquad K_2 = 1.6 \times 10^8$$

3. Qualitative analysis experiments reveal that AgCl dissolves in an excess of aqueous NH_3, whereas AgI does not. This result is due to the stability of $[Ag(NH_3)_2]^+$, which is sufficiently large to make the AgCl dissolve but not large enough to dissolve the less soluble AgI. A solution containing 0.15 M Cl^- and 0.15 M I^- is made 5 M in NH_3; then solid $AgNO_3$ is added in an amount equivalent to the $Cl^- + I^-$ concentrations. Calculate if AgCl and/or AgI will precipitate.

$$K_{sp}(AgCl) = 1.7 \times 10^{-10}$$
$$K_{sp}(AgI) = 8.5 \times 10^{-17}$$
$$\beta_2([Ag(NH_3)_2]^+) = 1.5 \times 10^7$$

4. (*a*) Calculate whether PbS would precipitate from a solution containing 0.5 M $EDTA^{4-}$, 0.001 M S^{2-}, and 0.01 M Pb^{2+}. (*b*) Make the same calculation for Ni^{2+}, Co^{2+}, Zn^{2+}, and Cd^{2+}.

Cation	$K\,[M\,(EDTA)]^{2-}$	$K_{sp}[M\,S]$
Pb^{2+}	2×10^{18}	4×10^{-26}
Ni^{2+}	3.6×10^{18}	1×10^{-22}
Co^{2+}	1.6×10^{16}	5×10^{-22}
Zn^{2+}	3.9×10^{16}	6×10^{-27}
Cd^{2+}	2.6×10^{16}	1×10^{-20}

References

Bjerrum, J. *Metal Ammine Formation in Aqueous Solution* (Haase, Copenhagen, 1941).
Rossotti, F. J. C. and Rossotti, H. *The Determination of Stability Constants* (McGraw-Hill, New York, 1961).
Fronaeus, S. "Determination of Stability Constants", in Jonassen, H. B. and Weissberger, A. (eds) *Technique of Inorganic Chemistry* (Wiley, New York, 1963).
Pearson, R. G. *Hard and Soft Acids and Bases* (Dowden, Hutchison and Ross, Stroudsburg, 1973).
    Data on the stability constants of metal complexes are continually reported in the scientific literature. A recent collection is presented in Martell, A. E and Smith, R. M. *Critical Stability Constants* (Plenum Press, New York, 1982).

Chapter 6

Kinetics and Mechanisms of Reactions of Coordination Compounds

One of the most important uses for metal complexes is in the homogeneous catalysis of reactions. Studies of metal enzymes (physiological catalysts) show that the site of reaction in the biological system is frequently a complexed metal ion. Many industrial processes depend directly on catalysis by metal complexes. The reaction of an alkene with carbon monoxide and hydrogen takes place in the presence of a cobalt complex, reaction (1).

$$CH_3CH{=}CH_2 + CO + H_2 \xrightarrow{\;[Co_2(CO)_8]\;} CH_3CH_2CH_2CHO \qquad (1)$$

This very important oxo reaction has been studied in detail; the catalyst is known to be $HCo(CO)_4$ which is generated in the reaction. The Monsanto Chemical Company recently developed a process for commercial production of acetic acid from CH_3OH and CO using a rhodium catalyst, reaction (2).

$$CH_3OH + CO \xrightarrow[180°C\ 35\ atm]{[RhI_2(CO)_2]^-} CH_3COOH \qquad (2)$$

Using a chiral rhodium complex as a catalyst Monsanto also manufactures 1-Dopa which is widely used for the treatment of Parkinson's disease.

These and many other applications of metal complexes are exciting the imagination of research chemists and increasing the production potential and versatility of our chemical industry. In order to exploit the use of metal complexes, we need to know more about some of the details of the reaction processes. This chapter illustrates the approach to such problems and provides examples of information obtained and reaction theories proposed.

In previous chapters, we have looked at a variety of reactions of coordination compounds. In Chapter 5, we found that the equilibrium constants for these reactions depend on the heat evolved and the amount of disorder produced (entropy). A favorable heat or entropy change is a necessary condition for the occurrence of a reaction. However, the reaction rate must also be sufficiently fast in order for a reaction to proceed. Reactions occur at a variety of speeds; some are immeasurably slow, and others are complete in a fraction of a second.

Some reactions, such as the very exothermic combination of H_2 and O_2 to give H_2O, do not occur until the mixture is ignited. Other, less exothermic reactions, such as the endothermic solution of salt in water, proceed rapidly at room temperature. This indicates that the rate of a reaction does not necessarily depend on the magnitude of the heat of the reaction. Reactions with very favorable equilibrium constants need not occur at a rapid rate. The rate of a chemical reaction is dependent on the nature of the path by which the reactants are transformed into products (*the mechanism of the reaction*). Knowing the mechanism of a reaction often makes it possible to understand the rate behavior of the reaction. What is more important in practice is that it is possible to learn a great deal about the reaction mechanism from the rate behavior of the reaction.

6.1   Rate of a reaction

The rate of a reaction such as reaction (3) can be expressed as the decrease in the number of moles of reactants, $[CoCl(NH_3)_5]^{2+}$ or H_2O, per second

$$[CoCl(NH_3)_5]^{2+} + H_2O \rightarrow [CoH_2O(NH_3)_5]^{3+} + Cl^- \qquad (3)$$

(or some other unit of time), or the increase in the number of moles of products, $[CoH_2O(NH_3)_5]^{3+}$ or Cl^- per second. Since the disappearance of 1 mole of $[CoCl(NH_3)_5]^{2+}$ produces 1 mole of $[CoH_2O(NH_3)_5]^{3+}$ and 1 mole of Cl^-, these three rates would be numerically equal. In general, the rate of any reaction can be defined as the change in concentration of any of the reactants or products per unit of time.

A convenient way of expressing a rate quantitatively is in terms of half-life. The *half-life* of a reaction is the amount of time necessary for half of the reactants to be consumed or the time for half of the products to be formed. The half-life of reaction (3) at 25°C is 113 h. This means that if one dissolved a salt containing $[CoCl(NH_3)_5]^{2+}$ in water at 25.0°C, after 113 h only half of the $[CoCl(NH_3)_5]^{2+}$ would remain, half having been converted to $[CoH_2O(NH_3)_5]^{3+}$ and Cl^- (Figure 6.1). After the next 113 h half of the remaining $[CoCl(NH_3)_5]^{2+}$ will have reacted, leaving only one-fourth of the original amount, and so on. Although H_2O is a reactant in this reaction, its concentration would not have been halved in the first 113 h, since as the solvent, it is present in very large excess. The concentrations of $[CoH_2O(NH_3)_5]^{3+}$ and Cl^- will be half of the values they will achieve when the reaction is complete.[1]

---

[1] The half-life concept is applicable in such a simple way only to first-order or pseudo-first-order processes, but this includes most reactions.

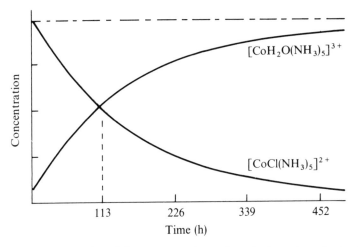

Figure 6.1   A plot of the concentrations of reactants and products of reaction (3) as a function of time at 25°C.

6.2   The rate law

Having defined the rate of a reaction, how can we learn something about rate phenomena from consideration of reaction mechanisms? The simplest type of reaction that one can visualize is the isomerization (4) or dissociation (5) of a molecule. Reactions of this stoichiometry can occur via complicated

$$A \rightarrow A' \tag{4}$$

$$A \rightarrow B + C \tag{5}$$

mechanisms in which a variety of intermediate products are formed; however, let us first choose the simplest mechanism, one in which A at some instant goes directly to A' (or B + C). One would expect, under these circumstances, that the rate of the reaction should depend on the concentration of A and no other species. The more molecules of A present, the greater will be the probability for one molecule to react. Thus the rate of reaction is directly proportional to the concentration of A, equation (6). This can be written by using a constant

$$\text{rate} \propto [A] \tag{6}$$

k that is called the *rate constant* and is a number characteristic of the reaction rate for a given temperature, equation (7).

$$\text{rate} = k[A] \tag{7}$$

For reactions that proceed rapidly, k is a large number; for slow reactions, k is small. There are a variety of reactions for which a simple rate expression of this type applies. An example is the conversion of *cis*-$[CoCl_2(en)_2]^+$ to *trans*-$[CoCl_2(en)_2]^+$ in methanol solution, reaction (8). The rate of conversion

$$cis\text{-}[CoCl_2(en)_2]^+ \xrightarrow{k} trans\text{-}[CoCl_2(en)_2]^+ \qquad (8)$$

of the *cis* to the *trans* isomer is equal to the product of the rate constant for the reaction and the concentration of the *cis* isomer, equation (9).

$$\text{rate} = k[cis\text{-}CoCl_2(en)_2]^+ \qquad (9)$$

The reaction may also proceed by a more complicated mechanism, equation (10). In this mechanism, A is converted to A' by a process involving the initial

$$A + D \xrightarrow{\text{slow}} E \quad (a) \qquad E \xrightarrow{\text{fast}} A' + D \quad (b) \qquad (10)$$

formation of an intermediate E (10a) and its subsequent conversion to A' (10b). The formation of E must involve a collision of A and D. The rate of this process should be proportional to the concentration of A and D, equation (11),

$$\text{rate of formation of } E \propto [A][D] \qquad (11)$$

since the probability of a collision depends directly on their concentrations. In a multistep process, the over-all reaction rate is determined by the slow step, which is called the *rate-determining step*. If the dissociation of E is much faster than its formation, A' will be formed effectively as rapidly as E. Then the rate of formation of A' is equal to the rate of formation of E. Rewriting the expression for E, we find the rate of formation of A' is given by expression (12). D is

$$\text{rate} = k[A][D] \qquad (12)$$

not consumed in the reaction, and yet the rate is dependent on its concentration; it is called a *catalyst*. The interconversion (13) of the optical isomers of $[Co(en)_3]^{3+}$ is catalyzed by $[Co(en)_3]^{2+}$, and the rate expression for the reaction has the form of expression (14). The rate-determining step in reaction (13) involves the transfer of an electron from $[Co(en)_3]^{2+}$ to $[Co(en)_3]^{3+}$ (Section 6.8).

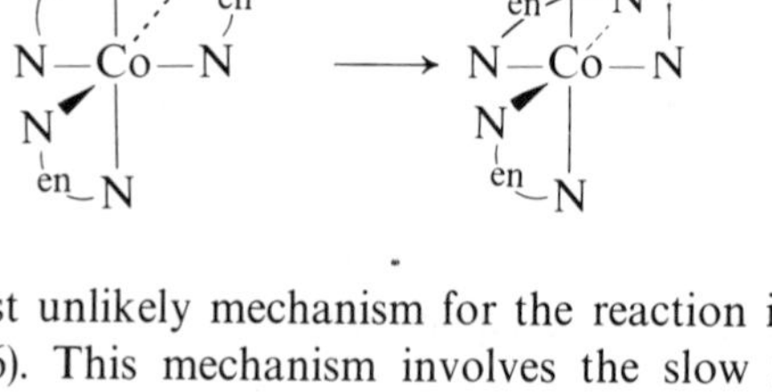

A third but most unlikely mechanism for the reaction is that shown by equations (15), (16). This mechanism involves the slow formation of an

$$\text{rate} = k[Co(en)_3^{3+}][Co(en)_3^{2+}] \qquad (14)$$

$$A + D + D \xrightarrow{\text{slow}} E \qquad (15)$$

$$E \xrightarrow{\text{fast}} A' + 2D \tag{16}$$

intermediate E by the collision of A and two molecules of D, equation (15). The rate of the formation of E, and of A', if E decomposes as rapidly as it is formed, is given by the expression (17). A collision of three bodies is very

$$\text{rate} = k[A][D][D] = k[A][D]^2 \tag{17}$$

improbable; hence, reactions that proceed by this type of process are very slow and very rare.

The rate expressions that were written for the three different paths from A to A' are called *rate laws*. They describe the effect of concentration on the rate of a reaction. A *first-order* rate law has the form of equation (7); the reaction rate (or simply the reaction) is said to have a *first-order dependence* on A or to be first order in [A]. Reactions that obey the *second-order rate law* (12), are said to have a first-order dependence on both [A] and [D]. The *third-order rate law* (17), indicates that reactions that obey this law will have a first-order rate dependence on [A] and a *second-order* dependence on [D].

The order of a reaction depends on the number of species and the number of times each species appears in the rate law. Often, the order of a reaction is equal to the number of particles that collide in the rate-determining step of the reaction. Later, however, we shall have examples in which the order of a reaction is less than the number of particles involved in the rate-determining step. From this discussion it should be apparent that the rate law for a reaction cannot be determined from stoichiometry. The rate law for the reaction A → A' may contain a variety of species not included in the chemical reaction; also, it need not include A or A'. If one can determine experimentally the rate law for a reaction, one may learn which species are involved in the rate-determining step and hence obtain vital information about the reaction mechanism.

6.3 Effective collisions

If it were possible to predict rate constants, then it would also be possible to determine which reactions would proceed at a rapid rate and which would be impractically slow. A theoretical approach to rate constants is available in collision theory. If each collision between reactants led to a reaction, the rate constant would be simply related to the probability for such collisions. In fact, in most reactions many collisions are ineffective; the rate constant is a measure of the collision effectiveness, and its magnitude stems primarily from the geometry and violence required in the collision.

The geometry of a collision must be suitable. For reactions of particles other than spherical molecules or ions, the particles must collide with a definite orientation for reaction to occur. For example, a cyanide ion must approach

reaction   no reaction   (18)

a metal ion with its carbon end in order for a metal–carbon bond to form, diagram (18). Reaction geometry is an important factor in gas-phase reactions, but it is less important in solution. Molecules in solution are effectively held in a cage by neighboring solvent molecules, and hence they normally collide with their neighbors many times before moving on to another site. Thus, whenever a CN^- comes near a metal ion, it will collide with it many times before it escapes from its cage; some of these collisions will almost certainly have the proper orientation for reaction.

The most important rate-determining factor in the majority of reactions is the *collision energy*. In the reaction of ammonia with aqueous Ag^+, the NH_3 molecule must take the place of a coordinated H_2O molecule. Collisions must provide the necessary energy for this process; otherwise the reaction does not occur. When molecules have used their collision energy to come to a configuration such that the reaction will proceed without further addition of energy, they are said to be in the *activated complex*. The amount of energy necessary to form the activated complex is called the *activation energy* (Figure 6.2). In reactions that have small activation energies, most collisions will be sufficiently energetic to produce a reaction. A very high activation energy will make all but the most violent collisions ineffective. The magnitude of the rate constant for a reaction, in general, inversely parallels the magnitude of the activation energy. The mechanism of the reaction determines the configuration and energy of the activated complex and, hence, the activation energy and rate of the reaction.

Reactions having large activation energies can be made to proceed at more convenient rates by increasing the reaction temperature or by using a catalyst. An increase in temperature increases the velocity of the reactant particles and, hence, the violence of their collisions. On the other hand, catalysts change the reaction mechanism so much that the new activated complex in which the catalyst is present can be formed by less energetic collisions.

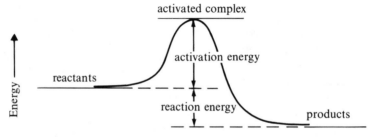

Figure 6.2 The relative energies of reactants, activated complex, and products of a reaction.

It is possible to visualize a variety of mechanisms for all reactions; the one that is observed is that which produces the fastest reaction under the conditions of the experiment. Other processes that are slower will make small or negligible contributions to the total reaction. Reaction mechanisms cannot be proven correct but can be proven wrong.

6.4   Inert and labile complexes

Complexes in which ligands are rapidly replaced by others are called *labile*; those in which ligand substitution is slow are called *inert*. To quantify this distinction, Henry Taube suggested that those complexes in which the substitution of ligands takes place in less than one minute be called labile. The reaction conditions are specified as a temperature of 25°C and 0.1 M reactants.

Although it is often found that a stable complex is inert and that an unstable complex is labile, no such correlation is required. Cyanide ion forms very stable complexes with metal ions such as Ni^{2+} and Hg^{2+}. The stability indicates that equilibrium (19) lies far to the right and that Ni^{2+} prefers CN^-

$$[Ni(H_2O)_6]^{2+} + 4CN^- \rightleftharpoons [Ni(CN)_4]^{2-} + 6H_2O \qquad (19)$$

to H_2O as a ligand. However, when ^{14}C-labeled cyanide ion is added to the solution, it is almost instantaneously incorporated into the complex,[1] equilibrium (20). Thus the stability of this complex does not ensure inertness.

$$[Ni(CN)_4]^{2-} + 4^{14}CN^- \rightleftharpoons [Ni(^{14}CN)_4]^{2-} + 4CN^- \qquad (20)$$

Cobalt(III) ammine complexes such as $[Co(NH_3)_6]^{3+}$ are unstable in acid solution. At equilibrium almost complete conversion to $[Co(H_2O)_6]^{2+}$, NH_4^+, and O_2 is observed, reaction (21). However, $[Co(NH_3)_6]^{3+}$ can be

$$4[Co(NH_3)_6]^{3+} + 20H^+ + 26H_2O \rightarrow 4[Co(H_2O)_6]^{2+} + 24NH_4^+ + O_2 \quad (21)$$

kept in acid solution for months at room temperature without noticeable decomposition. The rate of decomposition is very low; hence the compound is unstable in acid solution but inert.

In Chapter 5 the stability of coordination compounds was discussed; in the present chapter, reaction rate or lability is considered. Note that these terms relate to different phenomena. The stability of a complex depends on the difference in free energy between reactants and products. A stable compound will have a lower free energy than possible products.[2] The lability of a compound depends on the difference in free energy between the reactants and the activated complex; if this activation energy is large, the reaction will be slow.

---

[1] The labeled $^{14}CN^-$ is virtually identical chemically with unlabeled CN^-; therefore, the reaction proceeds until the ratio $^{14}CN^-/CN^-$ in the complex is equal to that in solution.

[2] The free energy of a substance is related to both its energy and entropy. Readers unfamiliar with the concept of free energy may think of this in terms of energy.

For six-coordinated complexes, it is possible to predict with some degree of reliability those which are labile and which are inert. Taube first called attention to this by pointing out that the electronic structure of a complex plays a significant role in the rate of its reactions. A classification of six-coordinated complexes in terms of the number and type of d electrons present in the central atom follows.

Labile complexes

(1) All complexes in which the central metal atom contains d electrons in e_g orbitals (the $d_{x^2-y^2}$ and x_{z^2} orbitals that point toward the six ligands; see Section 2.5), for example, $[Ga(C_2O_4)_3]^{3-}$, d^{10} $(t_{2g}{}^6e_g{}^4)$; $[Co(NH_3)_6]^{2+}$, $d^7(t_{2g}{}^5e_g{}^2)$; $[Cu(H_2O)_6]^{2+}$, $d^9(t_{2g}{}^6e_g{}^3)$; $[Ni(H_2O)_6]^{2+}$, $d^8(t_{2g}{}^6e_g{}^2)$; $[Fe(H_2O)_6]^{3+}$, d^5 $(t_{2g}{}^3e_g{}^2)$.
(2) All complexes that contain less than three d electrons, for example, $[Ti(H_2O)_6]^{3+}$, d^1; $[V(phen)_3]^{3+}$, d^2; $[CaEDTA]^{2-}$, d^0.

Inert complexes

Octahedral low-spin d^4, d^5, and d^6 systems, for example, $[Fe(CN)_6]^{3-}$, d^5 $(t_{2g}{}^5)$; $[Co(NO_2)_6]^{3-}$, d^6 $(t_{2g}{}^6)$; $[PtCl_6]^{2-}$, d^6 $(t_{2g}{}^6)$.

Octahedral d^3 complexes react relatively slowly and when the metal ion has an oxidation state of $+3$ or higher, their complexes are inert, for example $[Cr(H_2O)_6]^{3+}$ and other complexes of Cr^{3+}.

By using this classification, one can predict whether an octahedral complex is inert or labile is one knows its magnetic properties (whether it is high- or low-spin) and the number of d electrons present in the central atom.

The use of crystal field theory makes it possible to present a more detailed classification than simply *inert* and *labile*. The approach depends on a comparison of the CFSE of a coordination compound and of its activated complex (recall that "activated complex" refers to a configuration of reactant molecules such that the reaction can proceed without further addition of energy).

It is also possible to predict in more detail the rate behavior of complexes by considering the charge and size of their central atoms. The rules that were used to explain the stability of metal complexes (Section 5.2) often apply to kinetic behavior as well. Small, highly charged ions form the most stable complexes; likewise, these ions form complexes that react slowly. Thus, there is a decrease in lability with increasing charge of the central atom for the isoelectronic series $[AlF_6]^{3-} > [SiF_6]^{2-} > [PF_6]^- > SF_6$. Similarly, the rate of water exchange, equilibrium (22), decreases with increasing cationic charge

$$[M(H_2O)_6]^{n+} + 6H_2O^* \rightleftharpoons [M(H_2O^*)_6]^{n+} + 6H_2O \qquad (22)$$

in the order $[Na(H_2O)_n]^+ > [Mg(H_2O)_n]^{2+} > [Al(H_2O)_6]^{3+}$.

$$[M(H_2O)_x]^{n+} + xH_2O^* \rightleftharpoons [M(H_2O^*)_x]^{n+} + xH_2O$$

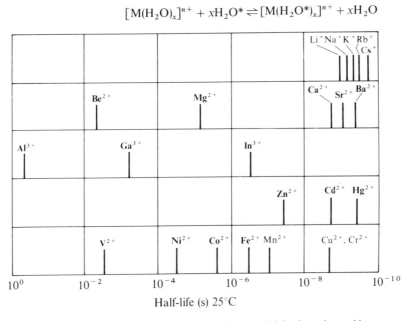

Figure 6.3   Half-lives for the exchange of water with hydrated metal ions. [Adapted and updated from Eigen, M. *Pure Appl. Chem.*, **6**, 105 (1963).]

Complexes having central atoms with small ionic radii react more slowly than those having larger central ions, for example $[Mg(H_2O)_6]^{2+} <$ $[Ca(H_2O)_6]^{2+} < [Sr(H_2O)_6]^{2+}$. For a series of octahedral metal complexes containing the same ligands, the complexes having central metal ions with the largest charge-to-radius ratios will react the slowest. The validity of this generalization is supported by the water exchange rate data summarized in Figure 6.3. Note that of the first-row transition elements in Figure 6.3, the slowest to react are $[V(H_2O)_6]^{2+}$ and $[Ni(H_2O)_6]^{2+}$, the d^3 and d^8 systems (which have the greatest crystal field stabilization energies). (The hydrated M^{2+} ions of the first-row transition elements are all high-spin complexes.) The rapid rate for $[Cu(H_2O)_6]^{2+}$ has been attributed to the exchange of water molecules above and below the square plane of the tetragonally distorted octahedral complex.

In general, four-coordinate complexes (both tetrahedral and square planar molecules) react more rapidly than analogous six-coordinate systems. As was stated earlier, the very stable $[Ni(CN)_4]^{2-}$ undergoes rapid exchange with $^{14}CN^-$, equilibrium (20). The rate of exchange is slow for six-coordinate complexes having about the same stability, for example $[Mn(CN)_6]^{4-}$ and $[Co(CN)_6]^{3-}$. The greater rapidity of reactions of four-coordinate complexes may be due to the fact that there is enough room around the central ion for a fifth group to enter the coordination sphere. The presence of an additional group would help to release one of the original ligands.

For square planar complexes it is not possible to apply successfully the charge-to-radius ratio generalization that works well for six-coordinate complexes. Thus, for the nickel triad, the size of the M^{2+} ions increases with increasing atomic number, but the rate of reaction decreases in the order $Ni^{2+} > Pd^{2+} \gg Pt^{2+}$. The rate of $*Cl^-$ exchange by $[AuCl_4]^-$ is approximately 10^4 times faster than that by $[PtCl_4]^{2-}$, although the reverse behavior is expected in terms of the charges on the metal ions.

As was noted earlier, the rate of a reaction depends upon the mechanism; this involves the configuration and energy of the activated complex and hence the activation energy. For octahedral systems the activation energy is strongly influenced by the breaking of metal–ligand bonds; therefore, a large positive charge on the central atom retards the loss of a ligand. In four-coordinate systems, the formation of new metal–ligand bonds is of increased importance, and it is favored by a large positive charge on the metal ion.

Therefore, the rules that predict rate behavior for six-coordinated systems will often not apply to complexes having smaller coordination numbers. Because rate behavior is dependent on mechanism and since reactions of metal complexes proceed by a variety of paths, it is impossible to make generalizations that apply to all complexes regardless of the type of mechanism by which they react. In spite of this, the rules that are outlined in this section are surprisingly consistent with the data on the rate behavior of octahedral complexes.

6.5   Mechanisms of substitution reactions

Now let us consider the application of kinetic and other techniques to the determination of reaction mechanisms. Reactions of coordination compounds can be divided into two broad categories: *substitution reactions* and *redox reactions*. In each of these, a variety of mechanistic paths may be operable.

Substitution reactions are generally classified as having either *dissociative* or *associative* mechanisms. We will use the general octahedral substitution reaction (23)

$$[MXA_5] + Y \rightarrow [MYA_5] + X \tag{23}$$

to illustrate these mechanisms. The extreme example of a dissociative mechanism, the D mechanism, equation (24),

$$\underset{\underset{A}{\overset{X}{|}}}{A-M-A} \xrightarrow[slow]{-X} \quad A-M-A \xrightarrow[fast]{+Y} \underset{\underset{A}{\overset{Y}{|}}}{A-M-A} \tag{24}$$

involves the rate-determining (slow) loss of X to give an intermediate with a coordination number diminished by one. Subsequent addition of Y to form the product is rapid.

The extreme example of an associative mechanism, the A mechanism, equation (25),

$$A-\underset{\underset{A}{|}}{\overset{\overset{X}{|}}{M}}-A \xrightarrow[\text{slow}]{+Y} \boxed{M} \xrightarrow[\text{fast}]{-X} A-\underset{\underset{A}{|}}{\overset{\overset{Y}{|}}{M}}-A \qquad (25)$$

involves the rate-determining (slow) addition of a ligand Y to form an intermediate with a coordination number increased by one. Then subsequent loss of X is fast. Both D and A mechanisms consist of two steps: one in which a bond is broken and one in which one is made.

Most reactions are intermediate in character; both bond breaking and bond making occur simultaneously. Such concerted processes are said to occur by an *I* (for interchange) mechanism. These processes are also divided into two classes. In reactions which have I_d mechanisms, the breaking of metal–ligand bonds has a greater affect on the rate and activation energy than does the formation of a new metal–ligand bond. In an I_a mechanism, bond making is more important.

The symbols S_N1 and S_N2 were introduced by Hughes and Ingold for organic reactions. The term S_N1 is used for dissociative mechanisms and means substitution nucleophilic unimolecular. The term S_N2 is used for mechanisms in which bond making is important and means substitution nucleophilic bimolecular. Let us now look at several systems to see how mechanistic information on substitution reactions of coordination compounds was obtained.

6.6   Octahedral substitution reactions

The most fundamental substitution reaction in aqueous solution, water exchange, reaction (22), has been studied for a variety of metal ions (Figure 6.3). Exchange of water in the coordination sphere of a metal with bulk solvent water occurs very rapidly for most metal ions, and therefore the rates of these reactions were studied primarily by relaxation techniques. In these methods, a system at equilibrium is disturbed, for example, by a very sudden increase in temperature. Under the new condition—higher temperature—the system will no longer be at equilibrium. The rate of equilibration can then be measured. If one can change the temperature of a solution in 10^{-8} s, then one can measure the rates of reactions that take longer than 10^{-8} s.

In Section 6.4, it was noted that the rate of water exchange decreases markedly as the charge on the metal ion increases and its size decreases. An increased charge-to-radius ratio is expected to strengthen metal–ligand bonds and hence retard metal–ligand bond cleavage. In contrast, an increase in the charge-to-radius ratio is expected to attract incoming ligands and aid associative reactions. Thus kinetic data on water exchange are consistent with a dissociative mechanism.

The rate of replacement of coordinated water by SO_4^{2-}, $S_2O_3^{2-}$, EDTA, and other species has also been measured for a variety of metal ions. It is believed that the mechanism for many of these reactions involves the very rapid association of the hydrated ion with the entering anion to form an ion pair (in which the anion is in the second coordination sphere of the metal ion), equilibrium (26). This ion pair then reacts to give the observed product in

$$[M(H_2O)_x]^{m+} + Y^{2-} \xrightarrow[\text{fast}]{\text{very}} [M(H_2O)_x]^{m+} \cdot Y^{2-} \qquad (26)$$

which the anion has replaced a water molecule, reaction (27).

$$[M(H_2O)_x]^{m+} \cdot Y^{2-} \xrightarrow{\text{slower}} [MY(H_2O)_{x-1}]^{((m-2)+} + H_2O \qquad (27)$$

For a variety of metal ions it has been found that the rate of reaction (27) is similar to its rate of water exchange; the nature of the entering ligand has little influence on the rate. Therefore, bond formation to the entering ligand apparently contributes little to the energetics of the rate-determining step. This and the fact that the rate is similar to water exchange implies that a similar I_d mechanism is involved for both water exchange and water substitution by an entering ligand.

Probably, the most widely studied coordination compounds are the ammine complexes of cobalt(III). Their stability, ease of preparation, and slow reactions makes them particularly amenable to kinetic study. Since work on these complexes has been done almost exclusively in water, the reactions of the complexes with the solvent water had to be considered first. In general, ammonia or amines coordinated to cobalt(III) are observed to be replaced so slowly by water that only the replacement of ligands other than amines is usually considered.

The rates of reactions of the type of reaction (28) have been studied and

$$[CoX(NH_3)_5]^{2+} + H_2O \rightarrow [CoOH_2(NH_3)_5]^{3+} + X^- \qquad (28)$$

found to be first order in the cobalt complex. (X can be any of a variety of anions.) Since in aqueous solution the concentration of H_2O is always about 55.5 M, the effect of changes in water concentration on the reaction rate cannot be determined. Rate laws (29) and (30) are experimentally indis-

$$\text{rate} = k[CoX(NH_3)_5^{2+}] \qquad (29)$$

$$\text{rate} = k'[CoX(NH_3)_5^{2+}][H_2O] \qquad (30)$$

tinguishable in aqueous solution, since k may simply be $k'[H_2O] = k'[55.5]$. Therefore, the rate law does not tell us whether H_2O is involved in the rate-determining step of the reaction. The decision as to whether these reactions proceed by displacement of X by H_2O or by dissociation followed by addition of H_2O must be made from other experimental data.

Two of the many types of experiments which have provided good mechanistic evidence are presented here. The rate of hydrolysis (replacement of one chloride by water) of *trans*-$[CoCl_2(NH_3)_4]^+$ is approximately 10^3 times faster than that of $[CoCl(NH_3)_5]^{2+}$. Since a decrease in rate is observed as the charge on the complex increases, a dissociative process seems to be operative.

Another piece of evidence results from the study of the hydrolyses of a series of complexes related to *trans*-$[CoCl_2(en)_2]^+$. In these complexes, the ethylenediamine is replaced by similar diamines in which H atoms on C were replaced by CH_3 groups. The complexes containing substituted diamines react more rapidly than the ethylenediamine complex. The replacement of H by CH_3 increases the bulk of the ligands. Models of these compounds show that this should make it more difficult for an attacking ligand to approach the metal atom. This steric crowding should retard an associative reaction. By crowding the vicinity of the metal atom with bulky ligands, one enhances a dissociative process, since removal of one ligand reduces congestion around the metal. The increase in rate observed when the more bulky ligands were used is good evidence for a dissociative mechanism.

As a result of a large number of studies on acido amine complexes of cobalt(III), it appears that replacement of the acido group by H_2O occurs by a process that is primarily dissociative in character. The ligand–cobalt bond must be stretched to some critical distance before a H_2O molecule begins to enter the coordination sphere.

Replacement of an acido group (X^-) in a cobalt(III) ammine with a group other than H_2O, reaction (31), has been observed to take place by initial

$$[CoX(NH_3)_5]^{2+} + Y^- \rightarrow [CoY(NH_3)_5]^{2+} + X^- \tag{31}$$

substitution by solvent H_2O with subsequent replacement of water by the new group Y, reaction (32). Therefore, in a number of cobalt(III) reactions

$$[CoX(NH_3)_5]^{2+} \xrightarrow[\text{slow}]{H_2O} [CoH_2O(NH_3)_5]^{3+} \xrightarrow[\text{fast}]{Y} [CoY(NH_3)_5]^{2+} \tag{32}$$

the rates of reaction (31) are the same as the rate of hydrolysis, reaction (28).

Hydroxide ion is different from other reagents with respect to its reactivity toward cobalt(III) ammine complexes. It reacts very rapidly (as much as 10^6 times faster than H_2O) with cobalt(III) ammine complexes in a *base hydrolysis* reaction (33). In this reaction, a first-order dependence on the substituting

$$[CoCl(NH_3)_5]^{2+} + OH^- \rightarrow [CoOH(NH_3)_5]^{2+} + Cl^- \tag{33}$$

ligand OH^- is observed, equation (34). The second-order kinetics and unusually rapid reaction suggest that OH^- is an exceptionally good nucleophilic

$$\text{rate} = k[\text{CoCl(NH}_3)_5{}^{2+}][\text{OH}^-] \tag{34}$$

reagent toward cobalt(III) and that reaction proceeds through an associative-type intermediate. However, an alternative mechanism, reactions (35) to (37),

$$[\text{Co(Cl(NH}_3)_5]^{2+} + \text{OH}^- \xrightarrow{\text{fast}} [\text{CoClNH}_2(\text{NH}_3)_4]^+ + \text{H}_2\text{O} \tag{35}$$

$$[\text{CoClNH}_2(\text{NH}_3)_4]^+ \xrightarrow{\text{slow}} [\text{CoNH}_2(\text{NH}_3)_4]^{2+} + \text{Cl}^- \tag{36}$$

$$[\text{CoNH}_2(\text{NH}_3)_4]^{2+} + \text{H}_2\text{O} \xrightarrow{\text{fast}} [\text{CoOH(NH}_3)_5]^{2+} \tag{37}$$

will also explain this behavior. In reaction (35), $[\text{CoCl(NH}_3)_5]^{2+}$ acts as a Brønsted acid to give $[\text{Co(NH}_3)_4\text{NH}_2\text{Cl}]^+$, which is known as an *amido* ($:\ddot{\text{N}}\text{H}_2{}^-$ containing) compound and which is the conjugate base of $[\text{CoCl(NH}_3)_5]^{2+}$. The reaction then proceeds by a dissociative process, reaction (36), to give a five-coordinated intermediate which reacts with solvent to give the observed product in reaction (37). This mechanism is consistent with second-order rate behavior, and yet it involves a dissociative mechanism. Since the reaction involves the conjugate base of the initial complex in an I_d rate-determining step, the term I_dCB is given to the mechanism.

The task of determining which mechanism best explains the experimental observations is difficult. However, there is convincing evidence to support the I_dCB hypothesis. When a reaction such as (33) is carried out in the presence of a relatively large concentration of a ligand Y^-, some $[\text{CoY(NH}_3)_5]^{2+}$ is formed. The amount formed relative to $[\text{CoOH(NH}_3)_5]^{2+}$ is independent of the OH^- concentration (in basic solution) and increases as the concentration of Y^- is raised. Direct displacement of a ligand by OH^- does not explain why Y^- should be incorporated, whereas a reactive five-coordinate intermediate as in reaction (36) would be expected to react with Y^- as well as H_2O. Such competition experiments show that an I_a mechanism is not correct for these reactions. The I_dCB mechanism is not proved to be true, but continues as a plausible mechanism. (It is not possible to prove that a mechanism is correct, but it is possible to prove one wrong.)

Another piece of evidence to support the I_dCB mechanism is that if there is no N–H hydrogen present in a cobalt(III) complex, the complex reacts slowly with OH^-. This suggests that acid–base properties of the complex are more important to the rate of reaction than are the nucleophilic properties of OH^-. This base hydrolysis reaction of cobalt(III) ammine complexes illustrates the fact that kinetic data often can be interpreted in more than one way and that rather subtle experiments must be performed to eliminate one or more possible mechanisms.

Substitution reactions of a wide variety of octahedral compounds have now been studied. Where mechanistic interpretation of the data has been made, a dissociative-type process has most frequently been postulated. This result should not be surprising, since six ligands around a central atom leave little room for adding another group. Nonetheless, evidence that entering groups do influence reaction rates is frequently found, and I_d mechanisms in which bond breaking is more important than bond making are common. There are also examples where bond making contributes significantly to the energetics, and therefore associative mechanisms cannot be discarded as paths for octahedral substitution.

6.7 Square planar substitution

Complexes in which the coordination number of the metal is less than six would be more likely to react by an associative process. Of complexes in which the coordination number of the metal is less than six, the four-coordinate platinum(II) complexes have been thoroughly studied. There is good experimental evidence for an I_a mechanism. The rates of reaction for some platinum(II) complexes having different charges are shown in Table 6.2. In the series of complexes in Table 6.2, the charge on the reactant platinum(II) complex goes from -2 to $+1$, and yet the rate changes by a factor of only two (quite a small effect). The breaking of a Pt—Cl bond should become more difficult as the charge on the complex becomes more positive; however, the formation of a new bond should become more favorable. The small effect of charge on the complex on reaction rate suggests that both bond making and bond breaking are important, as is characteristic in an I_a process.

Table 6.2 The rates of some reactions of platinum(II) complexes

Reaction	$t_{\frac{1}{2}}$ at 25°C, (min)
$[PtCl_4]^{2-} + H_2O \rightarrow [PtCl_3H_2O]^- + Cl^-$	300
$[PtCl_3NH_3]^- + H_2O \rightarrow [PtCl_2H_2ONH_3]^0 + Cl^-$	310
$cis\text{-}[PtCl_2(NH_3)_2]^0 + H_2O \rightarrow [PtClH_2O(NH_3)_2]^+ + Cl^-$	300
$[PtCl(NH_3)_3]^+ + H_2O \rightarrow [PtH_2O(NH_3)_3]^{2+} + Cl^-$	690

Good evidence for the importance of the entering ligand would be second-order kinetics, first order in platinum(II) complex and first order in entering ligand. This, in fact, was found for reactions of many different platinum(II) complexes with a variety of different ligands. There is a slight complication because the solvent water also behaves as a potential ligand. The result is that reactions such as (38) obey a two-term rate law, equation (39). This type of

$$[PtCl(NH_3)_3]^+ + Br^- \xrightarrow{\text{H}_2\text{O}} [PtBr(NH_3)_3]^+ + Cl^- \qquad (38)$$

$$\text{rate} = k[PtCl(NH_3)_3{}^+] + k'[PtCl(NH_3)_3{}^+][Br^-] \qquad (39)$$

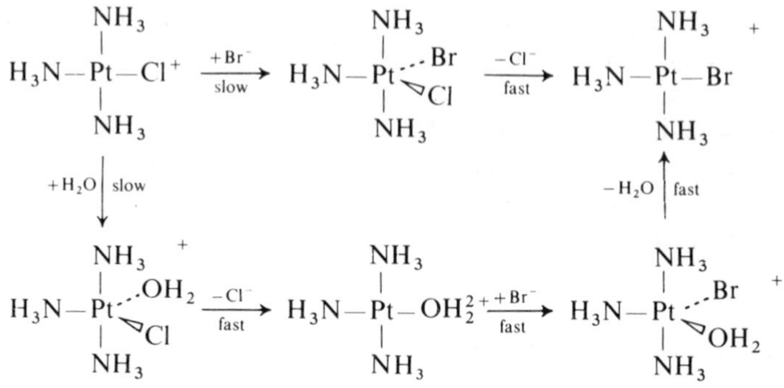

Figure 6.4   The mechanism for reaction (38).

rate law indicates the reaction is occurring by two mechanistic paths, only one of which involves Br^- in the rate-determining step.

Since experience with platinum(II) reactions suggests that the Br^- indepedent path is not an I_d process, it has been postulated that solvent H_2O replaces Cl^- in a slow step and is subsequently replaced in a rapid step by Br^-. The postulated mechanism is presented in Figure 6.4. That the solvent can participate in this way was demonstrated using experiments in which a similar reaction was carried out in a variety of different solvents. In solvents that are poor ligands (CCl_4, C_6H_6), only second-order kinetic behavior is observed and the entering ligand presumably enters the complex directly; in solvents that are good ligands (H_2O, alcohols), the first-order path also contributes to the reaction.

In substitution reactions of both square planar platinum(II) complexes and octahedral cobalt(III) complexes, the influence of solvent is quite marked. Note that in all solution processes the solvent plays an important role. Thus the behavior observed in H_2O may differ markedly from that found in other solvents.

It is widely accepted that square planar platinum(II) complexes react by an I_a mechanism. Kinetic work on substitution reactions of other square complexes suggests that I_a processes predominate in these systems. When the entering ligand plays a role in determining the rate of a reaction, it is important to learn which ligands promote the most rapid reactions.

Rate studies have indicated that ligands that exhibit a large *trans* effect (Section 4.8) also add rapidly to platinum(II) complexes. Groups such as phosphines, SCN^-, and I^- react rapidly with platinum(II) complexes; amines, Br^-, and Cl^- react at an intermediate rate; and H_2O and OH^- react slowly. This behavior reflects in part the *nucleophilicity* (the attraction toward a positive center) of these groups and indicates that OH^- is a poor nucleophile, toward platinum(II). However, the order of reactivity is clearly not an indication of only the attraction of ligands toward a positive center. If it were, Cl^- should react more rapidly than the larger anions Br^- and I^-. The observed order of reactivity can be related to the ease with which the entering ligand can release its electrons to platinum(II). Iodide ion, being less electronegative, is a better electron releasing agent than is Cl^-. A reasonably good correlation between the reactivity of the entering group and its oxidation potential is found. In general, the more easily a group is oxidized the more rapidly it reacts with platinum(II) complexes. The order of reactivity can also be related to whether the nucleophile is a hard or soft base (see p. 87). Platinum(II) is a soft acid and is observed both to form more stable complexes with soft bases and to react more rapidly with them.

Relatively few rate and mechanistic studies on tetrahedral complexes have been reported. These complexes are less common than octahedral complexes and their substitution reactions are frequently so fast that their study requires techniques such as stopped flow, temperature jump, or nuclear magnetic resonance. That the reactions are fast indicates that associative processes are occurring, since the activation energy for a reaction will be reduced when an incoming group can assist in the cleavage of a metal–ligand bond.

However, substitution reactions of some tetrahedral organometallic complexes such as $Ni(CO)_4$ are slow, and their rates are not dependent on the concentration of the entering ligand. This suggests that these reactions occur by a dissociative mechanism and has led to new insights into substitution mechanisms. As was noted in earlier chapters, complexes of transition metals with effective atomic numbers equal to the atomic number of a noble gas are said to obey the eighteen-electron rule and are relatively abundant and stable. The eighteen-electron rule works very well with organometallic systems and the following kinetic and mechanistic generalizations also work well with such compounds. The principles are generally useful, but there are many exceptions in the chemistry of classical Werner complexes.

Eighteen-electron complexes react more slowly than similar complexes with either more or less electrons. The eighteen-electron rule explains why some reactions are associative and others dissociative. Complexes in which the metal has sixteen or less "valence" electrons tend to react by associative mechanisms, since the metal has vacant low-energy orbitals which can be used to form a bond with the entering ligand. This orbital can accept an electron pair from an entering ligand and provide a path for associative substitution. Substitution reactions in square planar complexes illustrate this point, reaction (40).

$$[PtCl(NH_3)_3]^+ \xrightarrow[I_a]{+I^-} [PtClI(NH_3)_3] \xrightarrow[fast]{-Cl^-} [PtI(NH_3)_3]^+ \quad (40)$$
$$\text{16-electron} \qquad\qquad \text{18-electron} \qquad\qquad \text{16-electron}$$

Complexes in which the metal has eighteen or more valence electrons tend to react by dissociative mechanisms, because empty, low-energy orbitals are not available. For example, tetrahedral $Ni(CO)_4$ and trigonal bipyramidal $Fe(CO)_5$ react slowly with a nucleophile, reaction (41) and with no rate

$$Ni(CO)_4 \xrightarrow{-CO} Ni(CO)_3 \xrightarrow{+Nu} [Ni(CO)_3Nu] \qquad (41)$$

18-electron   16-electron   18-electron

dependence on the entering nucleophile. Both are eighteen-electron systems. Many stable organometallic compounds are either sixteen- or eighteen-electron systems and, therefore, one finds the sixteen-electron compounds reacting by associative mechanisms and the eighteen-electron compounds by dissociative ones.

It is interesting that two eighteen-electron complexes which are extremely similar to $Ni(CO)_4$, $Co(CO)_3NO$ and $Fe(CO)_2(NO)_2$, react rapidly with CO in a second-order process. This is explained by the ability of NO to accept a pair of electrons from the metal and in this way vacate a low-energy metal orbital suitable for accepting a pair of electrons from an entering ligand, reaction (42).

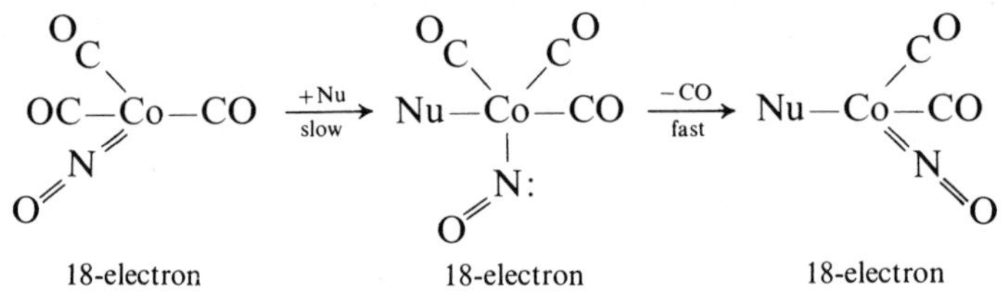

$$(42)$$

18-electron   18-electron   18-electron

These and other apparent exceptions to the eighteen-electron rule prompted an addition to the rule which states: substitution reactions of eighteen-electron transition metal compounds may proceed by an associative mechanism provided the metal can delocalize a pair of electrons onto one of its ligands.

6.8 Mechanisms for redox reactions

Let us now consider redox reactions, the other category of reactions of coordination compounds. Redox reactions are those in which the oxidation states of some atoms change. In reaction (43), the oxidation state of cobalt

$$[CoCl(NH_3)_5]^{2+} + [Cr(H_2O)_6]^{2+} + 5H_3O^+ \rightarrow$$
$$[Co(H_2O)_6]^{2+} + [CrCl(H_2O)_5]^{2+} + 5NH_4^+ \quad (43)$$

changes from $3+$ to $2+$ (cobalt is reduced); the oxidation state of chromium increases from $2+$ to $3+$ (chromium is oxidized). This change of oxidation state implies that an electron was transferred from chromium(II) to cobalt(III), reactions (44), (45). The mechanism for this reaction should suggest how the

$$Cr(II) \rightarrow Cr(III) + e \qquad (44)$$

$$e + Co(III) \rightarrow Co(II) \qquad (45)$$

electron is transferred. Two basic paths appear to be possible, In one, the electron effectively hops from one intact complex to the other: this is called the *outer-sphere* mechanism. In the other process, the oxidant and reductant are attached to each other by a bridging molecule, atom, or ion through which the electron can pass: this is called the *inner-sphere* mechanism.

Elegant experiments that demonstrated the validity of the inner-sphere path were performed by Taube and his coworkers. Reaction (43) was one of many studied. This test tube experiment conceived by Taube on the basis of his knowledge of substitution lability of metal complexes, was the seminal work which led to his Nobel prize. It was observed that in the reduction of $[CoCl(NH_3)_5]^{2+}$ by Cr^{2+} the chromium(III) product always contained a chloride ion. In more detailed studies, $[CoCl(NH_3)_5]^{2+}$ containing radioactive $^{36}Cl^-$, was dissolved in a solution containing Cr^{2+} and unlabeled Cl^-. After the reduction, which is very rapid, the product $[CrCl(H_2O)_5]^{2+}$ was examined and found to contain only labeled $^{36}Cl^-$. This proved that the cobalt complex was the only source of the Cl^- that was eventually found in the chromium(III) complex. To explain these results, a mechanism in which the activated complex contains Co and Cr atoms linked by a chloride ion was proposed, structure (I). The chloride bridge provides a good path between the two

I

metal atoms for electron transfer, much as a copper wire connecting two electrodes provides a path. Once an electron is transferred from chromium(II) to cobalt(III), the chromium(III) formed attracts the Cl^- more strongly than does cobalt(II), and, therefore, the Cl^- becomes part of the chromium(III) complex. An electron hop from the chromium complex to the cobalt complex followed by the transfer of the $^{36}Cl^-$ seems unlikely. If this were the path, then unlabeled Cl^- from solution would be as readily incorporated into the chromium(III) complex as is the $^{36}Cl^-$ attached to cobalt.

Reaction (43) and similar reactions were very clever choices for this study because cobalt(III) and chromium(III) complexes are inert, whereas chromium(II) and cobalt(II) complexes are labile. Thus, the rapid redox reaction is complete long before any substitution reactions begin to occur on the cobalt(III) and chromium(III) complexes. The lability of $[Cr(H_2O)_6]^{2+}$

allows the complex to lose a molecule of water rapidly and form the activated bridged intermediate, structure I. The results obtained demand a mechanism in which the coordinated chloride ion never escapes alone into the solution; for, in that case, appreciable quantities of $[Cr(H_2O)_6]^{3+}$ and unlabeled $[CrCl(H_2O)]^{2+}$ would be formed. A mechanism in which $^{36}Cl^-$ is attached to both Cr and Co during the electron transfer appears to fit the experimental data very well.

Reductions of a series of cobalt(III) complexes, $[CoX(NH_3)_5]^{2+}$, with chromium(II) solutions were studied. Transfer of the X^- group to chromium occurs when X^- is NCS^-, N_3^-, PO_4^{3-}, $C_2H_3O_2^-$, Cl^-, Br^-, and SO_4^{2-} (46). This suggests that all of these reactions are of the inner-sphere type.

$$[CoX(NH_3)_5]^{2+} + [Cr(H_2O)_6]^{2+} + 5H_3O^+ \rightarrow$$
$$[Co(H_2O)_6]^{2+} + [CrX(H_2O)_5]^{2+} + 5NH_4^+ \quad (46)$$

The rates of these reactions increased in the order $C_2H_3O_2^- < SO_4^{2-} < Cl^- < Br^-$. Presumably those ions that form bridges most readily and those that provide the best path for electrons produce the fastest reactions.

Redox reactions that proceed by electron transfer through a bridging group are quite common. In the reactions cited, transfer of the bridging atom accompanies the redox reaction. This is not a necessary result of the mechanism, but in its absence it is difficult to determine whether or not a bridging atom is involved in electron transfer.

There are a variety of redox reactions of metal complexes that probably occur by an outer-sphere mechanism. The rate of the redox reaction (47)

$$[*Fe(CN)_6]^{4-} + [Fe(CN)_6]^{3-} \rightarrow [*Fe(CN)_6]^{3-} + [Fe(CN)_6]^{4-} \quad (47)$$

(which is actually no reaction at all) can be studied by labeling either of the complexes with an isotope of Fe or C; the reaction is very rapid. Both iron(II)cyanide and iron(III)cyanide ions are substitution inert ($[Fe(CN)_6]^{4-}$ is a low-spin d^6 system; $[Fe(CN)_6]^{3-}$ is a low-spin d^5 system); therefore, loss or exchange of CN^-, or any substitution reaction, is very slow. The fact that the redox reaction is very fast whereas substitution reactions are very slow essentially eliminates the possibility of electron transfer through a bridged activated complex, since formation of the inner-sphere activated complex amounts to a substitution process.

When one rules out the bridge mechanism, one is left with an outer-sphere mechanism. On theoretical grounds, there is a critical requirement for such a process. The *Franck–Condon principle* states that there can be no appreciable change of atomic arrangement during the time of an electronic transition. That is, very light electrons move much more rapidly than the much heavier atoms. Let us consider the effect of this on an electron transfer process. The ligands are nearer the smaller Fe^{3+} ion than the larger Fe^{2+} (see structure II). During the transfer of an electron from $[Fe(CN)_6]^{4-}$ to $[Fe(CN)_6]^{3-}$, none of the Fe, C, or N atoms move. The electron transfer, therefore, results in the formation

$$
\begin{array}{c}
\text{CN} \\
\Big| \quad \text{CN} \\
\text{NC}\!-\!\!-\!\text{M}\!-\!\!-\!\text{CN} \\
\text{NC}^{\!\nearrow} \quad \Big| \\
\quad \text{CN}
\end{array}
\Bigg]^{4-}
\qquad
\begin{array}{c}
\text{CN} \\
\Big| \quad \text{CN} \\
\text{NC}\!-\!\text{M}\!-\!\text{CN} \\
\text{NC}^{\!\swarrow} \quad \Big| \\
\quad \text{CN}
\end{array}
\Bigg]^{3-}
$$

II

of an $[Fe(CN)_6]^{3-}$ in which the Fe—C bonds are too long and an $[Fe(CN)_6]^{4-}$ in which the Fe—C bonds are too short. Both of the products are of a higher energy than the normal ions in which the Fe—C bonds have their proper length (the length that gives the lowest energy).

The process described here is an example of a perpetual-motion machine. We took $[Fe(CN)_6]^{3-}$ and $[Fe(CN)_6]^{4-}$ ions, transferred an electron, and immediately obtained the same two species but with each now having an excess of energy. A process in which there is a net gain in energy cannot take place; hence this description of the reaction must be wrong. The reaction can occur only if we impart at least as much energy as we remove. Therefore, before electron transfer can occur, the Fe—C bonds in $[Fe(CN)_6]^{4-}$ must become shorter, the Fe—C bonds in $[Fe(CN)_6]^{3-}$ must become longer; and for this to be so, energy must be added to the system. For this reaction, a suitable configuration for reaction would be one in which the $[Fe(CN)_6]^{3-}$ and $[Fe(CN)_6]^{4-}$ ions have equivalent geometries. Then the products and reactants in the electron transfer process would be equivalent, and so no energy would be produced as a result of the electron transfer.

One can understand the rates of many direct electron transfer reactions by considering the amount of energy necessary to make the oxidant and reductant look alike. Since $[Fe(CN)_6]^{3-}$ and $[Fe(CN)_6]^{4-}$ are rather similar, a relatively small addition of energy (the activation energy) will make the ions alike; thus electron transfer can occur rapidly. Reaction (48) is very slow. The

$$[*Co(NH_3)_6]^{3+} + [Co(NH_3)_6]^{2+} \rightarrow [*Co(NH_3)_6]^{2+} + [Co(NH_3)_6]^{3+} \quad (48)$$

complexes $[Co(NH_3)_6]^{2+}$ and $[Co(NH_3)_6]^{3+}$ do not differ greatly in size, and hence one might expect electron exchange between these two complexes to be rapid. The two complexes do, however, differ in electronic configuration; the $[Co(NH_3)_6]^{2+}$ is $t_{2g}^5 e_g^2$, $[Co(NH_3)_6]^{3+}$ is t_{2g}^6. Hence, both the length of the Co—N bonds and the electronic configurations must change prior to electron transfer. This is the reason for the very slow reaction.

Other factors influence the rate of direct electron transfer processes. For example, the greater the conductivity of their ligands the more readily should electron transfer proceed between two complexes. Multiple bonds in cyanide ions are expected to provide a good path for electrons, and, indeed, electron transfer between a variety of similar cyanide complexes has been found to be rapid. The same is true of the electron conducting aromatic ligands in $[M(phen)_3]^{n+}$ and $[M(bipy)_3]^{n+}$ relative to the saturated ligands in $[M(en)_3]^{n+}$ and $[M(NH_3)_6]^{n+}$.

Problems

1. Designate whether the following complexes are expected to be inert or labile and give a reason for your choice.

$[Al(C_2O_4)_3]^{3-}$ $[V(H_2O)_6]^{3+}$
$[Cr(C_2O_4)_3]^{3-}$ $[V(H_2O)_6]^{2+}$
$[CoF_6]^{3-}$ (high-spin) $[Ni(NH_3)_6]^{2+}$
$[Fe(CN)_6]^{4-}$ (low-spin) $[PtCl_6]^{2-}$ (low-spin)

2. For analogous complexes of each of the following series of metal ions, indicate the order of decreasing lability and explain your answer. (*a*) Mg^{2+}, Ca^{2+}, Sr^{2+}, Ba^{2+}, and Ra^{2+}. (*b*) Mg^{2+}, Al^{3+}, and Si^{4+}. (*c*) High-spin Ca^{2+}, V^{2+}, Cr^{2+}, Mn^{2+}, Fe^{2+}, Co^{2+}, Ni^{2+}, Cu^{2+}, and Zn^{2+}.

3. Explain why $[Co(NH_3)_6]^{3+}$ is reduced much more slowly than $[CoCl(NH_3)_5]^{2+}$ by $[Cr(H_2O)_6]^{2+}$. Write the formula of the chromium(III) reaction product in each case.

4. The complex $[CoCO_3(NH_3)_5]^+$ will react in acid solution to generate CO_2. One might expect the rate law for this reaction to have the form

$$rate = k[CoCO_3(NH_3)_5{}^+][H^+]$$

$$[CoOCO_2(NH_3)_5]^+ + 2H^+ \xrightarrow{\ H_2{}^{18}O\ } [CoOH_2(NH_3)_5]^{3+} + CO_2$$

It is probable that the reaction will occur without the incorporation of ^{18}O into the products from labeled solvent $H_2{}^{18}O$. Propose a reasonable mechanism that is consistent with this rate law and the absence of labeled ^{18}O in the products.

5. Both of the complexes $[AuCl_4]^-$ and $[AuCl(dien)]^{2+}$ undergo very rapid radiochloride exchange:

$$[AuCl(dien)]^{2+} + {}^*Cl^- \rightarrow [Au^*Cl(dien)]^{2+} + Cl^-$$

The rate laws for these reactions have the form

$$rate = k[complex] + k'[complex][Cl^-]$$

Propose a reasonable mechanism that is consistent with these data.

6. The $[M(bipy)_3]^{2+}$ complexes of the first-row transition metals dissociate in water. The rates of these reactions have been measured and found to

$$[M(bipy)_3]^{2+} + 2H_2O \rightarrow [M(bipy)_2(H_2O)_2]^{2+} + bipy$$

increase in the order $Fe^{2+} < Ni^{2+} < Co^{2+} \sim Mn^{2+} \sim Cu^{2+} \sim Zn^{2+}$. Propose a reason for this order of reactivity.

7. On the basis of crystal field theory octahedral d^3 and d^8 systems are expected to react at comparable rates. Yet $[Cr(H_2O)_6]^{3+}$ is inert and $[Ni(H_2O)_6]^{2+}$ is labile (half-lives for water exchange are 40 h and 3×10^{-5} s, respectively). Account for this difference in rate behavior.

8. Important evidence cited to support I_dCB mechanisms for base hydrolysis of $[CoX(NH_3)_5]^{2+}$ systems is a study involving competition between OH- and Y-ligands (see p. 110). Explain how these data are consistent with the I_dCB mechanism and inconsistent with the I_a mechanism. Suggest related experiments which might help one decide between these mechanisms. (See Buckingham, D. A., Olsen, I. I. and Sargeson, A. M. *J. Am. Chem. Soc.*, **88**, 5443 (1966).)

References

Basolo, F. and Pearson, R. G. *Mechanisms of Inorganic Reactions*, 2nd edn (Wiley, New York, 1967).

Benson, D. *Mechanisms of Inorganic Reactions in Solution: An Introduction* (McGraw-Hill, New York, 1968).

Edwards, J. O. (ed.) *Inorganic Reaction Mechanisms* (Wiley, New York, 1970).

Langford, C. H. and Gray, H. B. *Ligand Substitution Processes* (Benjamin, New York, 1966).

Sykes, A. G. (ed.) *Advances in Inorganic and Bioinorganic Mechanisms* (Academic, New York, 1982).

Taube, H. *Electron Transfer Reactions of Complex Ions in Solution* (Academic, New York, 1970).

Tobe, M. L. *Inorganic Reaction Mechanisms* (Thames and Nelson, London, 1972).

Wilkins, R. G. *The Study of Kinetics and Mechanism of Reactions of Transition Metal Complexes* (Allyn and Bacon, Boston, 1974).

Chapter 7

Organometallic Chemistry and Current Research Topics

Inorganic chemistry is experiencing a period of intensely exciting research activity for many different reasons. Advancements in theory and modern sophisticated instruments make it possible to understand better complicated inorganic systems. The demand for new materials by our technological world requires us to look beyond organic compounds. In this chapter, we will first introduce metal carbonyls and organometallic compounds and some basic ways in which they are made. Organometallic synthesis continues to be a very active area. We will then mention three areas of research activity in inorganic chemistry: (1) organometallic reactions and homogeneous catalysis, (2) bioinorganic chemistry, and (3) solid-state chemistry.

7.1 Preparation of metal carbonyls and organometallic compounds

Compounds containing transition metal–carbon bonds have been known for many years. The dye prussian blue ($Fe[Fe_2(CN)_6]_3$), which contains Fe—CN bonds, was perhaps the first example of such a coordination compound. The metal carbonyls $Ni(CO)_4$ and $Fe(CO)_5$ were prepared by the French chemist Mond in about 1890. The discovery of ferrocene, in 1951, is generally said to mark the beginning of modern transition metal organometallic chemistry. Since 1951, a wide variety of compounds containing transition metal–carbon bonds have been synthesized. These compounds include alkyl (for example, $CH_3Mn(CO)_5$) and aryl (for example, $[Pt\{P(C_2H_5)^3\}_2$-$(C_6H_5)_2]$) compounds in which metal–carbon bonds (σ bonds) exist. A variety of alkene compounds (structure I) are known in which the bonding can be described as involving sharing of π electrons of the alkene (an organic molecule containing double bonds) with the metal. The so-called sandwich compounds (structure II) in which a metal is located between two planar (or nearly so) cyclic carbon compounds, are a third type of organometallic compound. In the latter compounds, the bonding may be considered to involve the sharing of π electrons of the organic molecule with d or hybrid orbitals on the metal.

Metal atoms in compounds containing carbon–metal bonds frequently

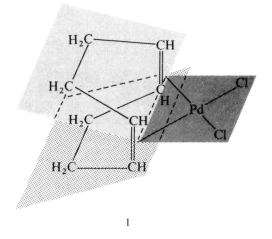

have formal oxidation states that are abnormally low; for example, metals in most metal carbonyls are formally zero valent. This fact and their molecular nature suggest that metal carbonyls might normally be prepared under reducing conditions, frequently in nonaqueous solvents. Many of these syntheses are performed in diglyme $[(CH_3OCH_2CH_2)_2O]$, tetrahydrofuran $(CH_2CH_2CH_2CH_2O)$, or diethyl ether, solvents in which the reactants and products are soluble and which are more resitant to reduction than water.

Preparation of metal carbonyls

Metal carbonyls were first prepared by Mond by the direct reaction of gaseous carbon monoxide with finely divided metals. Iron, cobalt, and nickel carbonyls can be prepared in this way, reaction (1). The formation of $Ni(CO)_4$

$$Ni + 4CO \xrightarrow[25°C]{1\ atm} \underset{colorless}{Ni(CO)_4} \quad (bp\ 43°C) \tag{1}$$

proceeds rapidly at room temperature and 1 atmosphere of CO. Higher pressures and temperatures are required to prepare $Co_2(CO)_8$. The Mond process for the metallurgy of nickel and cobalt separates these metals by the formation of $Ni(CO)_4$ at moderate temperatures and pressures and its subsequent decomposition to the metal and CO by stronger heating. Since $Co_2(CO)_8$ is formed only very slowly under the same conditions, and has a relatively low volatility, cobalt is left behind when gaseous $Ni(CO)_4$ is removed.

A variety of metal carbonyls are known. The eighteen-electron rule and Sidgwick's effective atomic number rule (Section 2.3) are very successful in explaining their stoichiometries. Simple monomeric carbonyls are expected for transition metals with even atomic numbers: $Cr(CO)_6$, $Fe(CO)_5$, $Ni(CO)_4$. The heavier members of the Cr and Fe families also form monomeric carbonyls of the predicted stoichiometry.

Transition metals with odd atomic numbers cannot achieve the expected EAN in monomeric compounds. In cases where carbonyls of these elements are known, the compounds contain more than one metal atom and metal–metal bonds; these effectively contribute one extra electron to each metal: $Mn_2(CO)_{10}$, $Co_2(CO)_8$. Many polynuclear metal carbonyls are known. The largest metal cluster that has been characterized is a polynuclear platinum carbonyl anion, $[Pt_{38}(CO)_{44}H_x]^{2-}$.

In 1959, $V(CO)_6$ was prepared. It is a black paramagnetic solid that decomposes at 70°C. It is the only stable metal carbonyl containing one metal atom that does not obey the eighteen-electron rule. It is a seventeen-electron system which is readily reduced to the stable eighteen-electron anion $[V(CO)_6]^-$, reaction (2). The seventeen-electron $V(CO)_6$ is very reactive; for example, it

$$\underset{black}{V(CO)_6} + Na \xrightarrow[diglyme]{} \underset{yellow}{[V(CO)_6]^-} + Na^+ \tag{2}$$

undergoes CO substitution reactions 10^{10} times faster than does the eighteen-electron $Cr(CO)_6$.

Metal carbonyls are commonly made by reduction of metal salts in the presence of CO under high pressure. A variety of reducing agents have been found to be effective. In reaction (3), sodium reduction produces a solvated sodium salt of $V(CO)_6^-$; this is subsequently oxidized by H^+ to $V(CO)_6$, reaction (4).

$$\underset{pink}{VCl_3} + 6Na + 6CO \xrightarrow[\substack{diglyme \\ 200\,atm}]{100°C} \underset{yellow}{[(diglyme)_2Na]^+[V(CO)_6]^-} \tag{3}$$

$$\underset{yellow}{[(diglyme)_2Na]^+[V(CO)_6]^-} \xrightarrow[H_3PO_4(100\%)]{25°C} \underset{black}{V(CO)_6} + H_2 \tag{4}$$

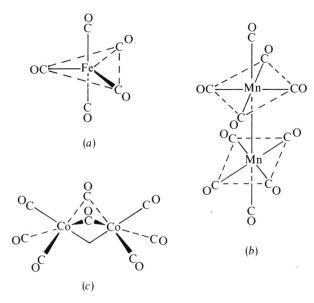

Figure 7.1 The structures of (*a*) Fe(CO)$_5$, (*b*) Mn$_2$(CO)$_{10}$, and (*c*) Co$_2$(CO)$_8$. Although this is the structure in the solid state, in solution the CO bridged form is in rapid equilibrium with (OC)$_4$Co–Co(CO)$_4$.

Other active metals, such as aluminium, as well as organic salts of active metals, for example C$_2$H$_5$MgBr and C$_6$H$_5$Li, are often used. In a few cases, metal carbonyls are conveniently prepared by reduction with the readily synthesized Fe(CO)$_5$, reaction (5). Carbon monoxide itself is an excellent

$$\underset{\text{blue}}{\text{WCl}_6} + 3\text{Fe(CO)}_5 \xrightarrow[\text{ether}]{90°C} \underset{\text{colorless}}{\text{W(CO)}_6} + 3\text{FeCl}_2 + 9\text{CO} \qquad (5)$$

reducing agent and, in some cases, serves this function as well as that of a ligand, reaction (6).

$$\underset{\text{yellow-brown}}{\text{Re}_2\text{O}_7} + 17\text{CO} \xrightarrow[\substack{\text{and pressure} \\ \text{of CO}}]{\text{heat}} \underset{\text{colorless}}{\text{Re}_2(\text{CO})_{10}} + 7\text{CO}_2 \qquad (6)$$

A variety of compounds can be obtained by reactions of metal carbonyls. Reactions characteristic of these compounds are illustrated in Figure 7.2 by some reactions of Fe(CO)$_5$.

Many simple metal carbonyls are volatile and very toxic: their harmful effects are cumulative, so exposure to them should be avoided.

Preparation of transition-metal–alkene compounds

In 1827, W. C. Zeise, a Danish pharmacist, found that ethene, C$_2$H$_4$, reacts with [PtCl$_4$]$^{2-}$, in dilute HCl solution to yield compounds containing both

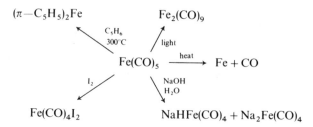

Figure 7.2 Reactions of Fe(CO)$_5$ illustrating reactions of metal carbonyls.

platinum and ethene. Structures of the reaction products, III and IV, were determined over 100 years later. These compounds can be described as square planar with the planar ethene bonds pointing between the two carbon atoms.[1] The salt K[PtCl$_3$(C$_2$H$_4$)] is called Zeise's salt.

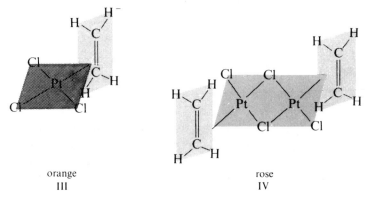

orange   rose
III   IV

Molecular orbital theoretical treatments of ethene (and other alkene) metal complexes describe the bonding as involving an overlap of an empty metal orbital with a filled π molecular orbital that is delocalized over the ethene molecule. Additional stability is provided by possible π bonding between suitably oriented filled metal orbitals and vacant antibonding alkene molecular orbitals (Figure 7.3).

In recent years, a wide variety of alkene and related compounds have been prepared. The most stable of these seem to be formed by molecules containing two double bonds that are so located that they can both bond to the same metal. One molecule of this type is cyclooctadiene, reaction (7). Alkene compounds are normally prepared by the direct reaction of an alkene with a metal salt or complex.

Preparation of sandwich compounds

Since 1950, a large number of transition-metal compounds have been prepared in which the metal atom has been described as the "meat" between

---

[1] In solution, the ethene can rotate freely around the bond it forms with platinum.

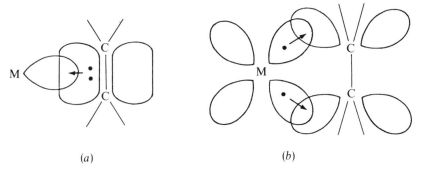

Figure 7.3 A representation of bonding in metal alkene complexes: (*a*) the σ bond in which a π MO on the alkene overlaps with a metal orbital; (*b*) the π bond in which the π^* antibonding MO of the alkene overlaps with a metal *d* orbital.

two flat organic molecules that are the "slices of bread" in the "sandwich" molecule. A variety of metals and organic molecules have been used successfully; the most stable compounds contain the cyclopentadiene anion, $C_5H_5^-$, structure V.

The first compound of this type to be recognized was ferrocene [bis(π-cyclopentadienyl)iron(II)], structure II. It is an orange crystalline compound that boils without decomposition at 249°C and is unaffected by aqueous NaOH or concentrated HCl. The compound is diamagnetic and nonpolar. On the basis of such chemical and physical properties, the sandwich structure II was proposed. This structure has been confirmed by X-ray diffraction studies. The 1973 Nobel prize in chemistry was awarded to Professor Ernst Otto Fischer of the Technische Universität in Munich, Germany and to Professor Geoffrey Wilkinson of Imperial College in London, England for their pioneering research on "metallocenes". They extended the discovery of ferrocene to the syntheses and reactions of other metallocenes and opened up a new area of chemistry with important bonding concepts and practical applications.

Most stable sandwich and alkene compounds are eighteen-electron systems. The $C_5H_5^-$ ion is treated as a six-electron donor, as is the benzene molecule; ethene is a two-electron donor. Combinations of ligands that donate the proper number of electrons to a metal to give it eighteen "valence" electrons produce stable compounds, for example $Fe(C_5H_5)_2$, $Mn(C_5H_5)(C_6H_6)$, and $Cr(C_6H_6)_2$.

In the ferrocene molecule, the cyclopentadiene anion reacts like an aromatic organic molecule. Because ferrocene is quite stable, it has been possible to perform reactions characteristic of an aromatic system on the ring without destroying the bonding to the metal, reaction (8). An extensive organic

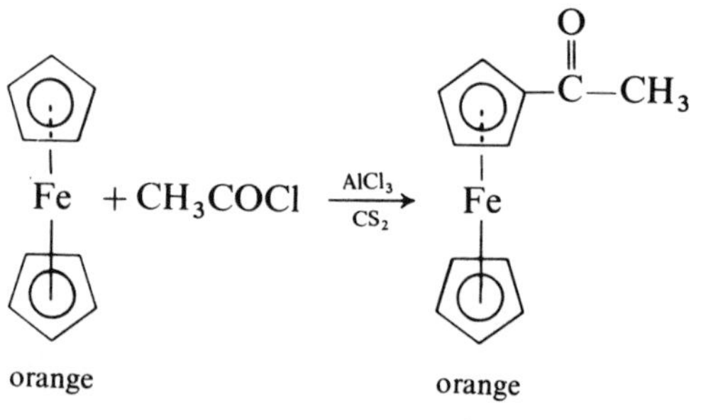

$$\text{orange} \qquad\qquad \text{orange} \tag{8}$$

chemistry of ferrocene, with a corresponding production of new ferrocene derivatives, has been developed.

Cyclopentadiene sandwich compounds are often prepared by reactions of metal halides with sodium cyclopentadienide, reaction (9). Dibenzene-

$$\underset{\text{green}}{FeCl_2} + 2C_5H_5Na \xrightarrow{\text{ether}} \underset{\text{orange}}{Fe(C_5H_5)_2} + 2NaCl \tag{9}$$

chromium, $Cr(C_6H_6)_2$, is best prepared by the reaction scheme (10). An intermediate chromium(I) compound is produced; it is then treated with a reducing

$$3CrCl_3 + 2Al + AlCl_3 + 6C_6H_6 \rightarrow 3[Cr(C_6H_6)_2]^+[AlCl_4]^-$$
violet

$$\xrightarrow[OH^-]{S_2O_4{}^{2-}} Cr(C_6H_6)_2 + SO_3{}^{2-} \qquad (10)$$
brown

agent such as dithionite ion $(S_2O_4{}^{2-})$. Dibenzene chromium and other benzene sandwich compounds are considerably less stable than many of the cyclopentadiene compounds. They are much more readily oxidized and decomposed under most reaction conditions.

Preparation of transition metal–carbon σ-bonded compounds

Transition metal compounds containing σ-bonded alkyl or aryl groups were rare until the 1960's. It has been found that the presence of ligands such as CO, $C_5H_5{}^-$, or PR_3 on transition metals greatly enhances the ability of transition metals to form σ-bonded organometallic compounds. Transition metal–carbon σ bonds are often produced by metathesis reactions in which one product is the organometallic compound and the other is a simple salt, reactions (11), (12). Transition metal fluorocarbon derivatives have also been

$$NaMn(CO)_5 + CH_3I \rightarrow CH_3Mn(CO)_5 + NaI \qquad (11)$$
colorless   colorless

$$cis\text{-}[PtBr_2\{P(C_2H_5)_3\}_2] + 2CH_3MgBr \xrightarrow{ether} \qquad (12)$$
colorless

$$cis\text{-}[Pt(CH_3)_2\{P(C_2H_5)_3\}_2] + 2MgBr_2$$
colorless

prepared and they are more stable than the corresponding hydrocarbon compounds, reactions (13), (14).

$$HMn(CO)_5 + C_2F_4 \xrightarrow[25°C]{pentane} MnHCF_2CF_2(CO)_5 \qquad (13)$$
colorless   colorless

$$Fe(CO)_5 + F_3CCF_2I \xrightarrow[45°C]{benzene} Fe(F_3CCF_2)(CO)_4I \qquad (14)$$
yellow   purple

A variety of organometallic compounds can be prepared by metal atom reactions. Atoms of a metal are vaporized from a very hot metal sample into an evacuated chamber. They are condensed onto a surface cooled with liquid nitrogen ($-196°C$) along with the desired ligand. Molecules which are too unstable to be prepared by more conventional techniques can be made in this way as well as stable molecules such as ferrocene and dibenzenechromium. Although solid chromium does not react with either liquid or gaseous C_6H_6, reaction (15), gaseous chromium atoms react even at low temperatures,

$$Cr(s) + C_6H_6(g) \xrightarrow{50°C} No \text{ reaction} \qquad (15)$$

reaction (16). This is readily understood when one recognizes that in reaction

$$Cr(g) + 2C_6H_6(g) \xrightarrow{-196\,C} Cr(C_6H_6)_2(s) \tag{16}$$

(15) chromium atoms have to be removed from solid chromium in order to form a chromium product. This requires a large amount of energy and makes the reaction nonspontaneous. By contrast, in reaction (16) the reactive chromium atoms are already formed and react rapidly with benzene and many other molecules.

More complicated reactions can also be achieved using metal atoms. Reaction (17) illustrates the formation of a metal alkyl complex.

$$Pd(g) + 2P(C_2H_5)_3(g) + CF_3Br \xrightarrow{-196\,C} solid$$

$$\xrightarrow{25\,C} [Pd(CF_3)Br\{P(C_2H_5)_3\}_2] \tag{17}$$

7.2 Organometallic reactions and homogeneous catalysis

In Chapter 4, reactions of coordination compounds were discussed and in Chapter 6, their mechanisms were introduced. The principles of those chapters also apply to organometallic chemistry; here we emphasize certain types of reactions which are of specific importance in organometallic chemistry, and particularly in their application to homogeneous catalysis. This differs from the more traditional *heterogeneous catalysis*, where the catalyst is a solid and reactants are liquids and/or gases. In *homogeneous catalysis*, the catalyst and reactants are in the same phase (usually dissolved in a liquid). Such homogeneous catalysis processes are becoming increasingly important in industry (see p. 97).

(1) *Addition reactions.* Certain square planar d^8 (sixteen-electron) complexes such as $[RhH(CO)L_2]$ play a particularly important role in catalysis.[1] These are sixteen-electron systems, having two less electrons than needed to complete a noble gas electronic configuration. It should not be surprising that such complexes add one ligand to become eighteen-electron complexes, reaction (18).

$$[RhH(CO)L_2] + L \rightarrow [RhH(CO)L_3] \tag{18}$$

The reverse reaction to reform the square planar complex is also well known. In contrast, it is generally not possible to isolate complexes made by adding a fifth ligand to traditional square planar complexes, such as $[Pt(NH_3)_4]^{2+}$.

---

[1] The letter L will be used to represent the ligand $P(C_6H_5)_3$, which is used frequently in organometallic complexes that are catalysts.

However, such species are thought to be intermediates in substitution reactions.

(2) *Oxidative–addition reactions.* In Section 4.10, oxidative addition reactions were introduced with examples in which a halogen, Br_2, was added to square planar d^8 systems. In such reactions, both oxidation and addition occur. Important reactions of this type in organometallic chemistry involve addition of molecules such as H_2 and organic halides which are not normally considered to be oxidizing agents, reactions (19) and (20).

$$[RhClL_3] + H_2 \rightarrow [RhClH_2L_3] \tag{19}$$

$$[RhI_2(CO)_2]^- + CH_3I \rightarrow [RhI_3CH_3(CO)_2]^- \tag{20}$$

In a formal sense, these are oxidation reactions, because rhodium(I) is oxidized to rhodium(III). These are also addition reactions because two ligands are added to square planar sixteen-electron systems which are transformed into octahedral eighteen-electron complexes. These reactions can also be viewed as an insertion reaction (see (4) below) in which a metal atom is inserted into a bond between two nonmetals.

(3) *Reductive elimination.* This is the reverse of oxidative addition. An example is reaction (21) which occurs in the catalytic hydrogenation of alkenes. This

$$\begin{array}{c}
H \quad \text{solvent} \\
| \\
L-Rh-CH_2CH_3 \\
| \\
L \quad Cl
\end{array}
\longrightarrow CH_3CH_3 +
\begin{array}{c}
L \\
| \\
L-Rh-\text{solvent} \\
| \\
Cl
\end{array}
\tag{21}$$

final step in the hydrogenation cycle produces the alkane reaction product and regenerates the square planar rhodium catalyst.

(4) *Insertion or ligand migration.* Notice that in reaction (22) the overall result

$$[Pt(H)ClL_2] + CH_2{=}CH_2 \rightarrow [Pt(CH_2CH_3)ClL_2] \tag{22}$$

is that ethene, an organic molecule, is inserted into a metal–hydrogen bond. However, physically this is not what happens. Instead, the alkene first adds to the metal; then the H^- ligand migrates onto a carbon while the other carbon forms a bond to the metal, reaction (23).

$$[L_2ClPt-H] \xrightarrow{CH_2=CH_2} \left[L_2ClPt\underset{H}{\overset{CH_2}{\underset{\shortparallel}{\diagup}}}CH_2 \right] \longrightarrow [L_2ClPt-CH_2CH_3] \tag{23}$$

Studies on reaction (24) using labeled CO demonstrate that the CO "inserted"

$$[CH_3-Mn(CO)_5] + CO \rightarrow [CH_3CO-Mn(CO)_5] \tag{24}$$

into the Mn–C bond was a CO already present in the complex and not a CO from solution. It appears that the CH$_3$ group migrates onto a coordinated CO, and a CO from solution fills the vacated coordination site. There is considerable evidence supporting ligand migration as the actual mechanism for such insertion reactions.

The reverse decarbonylation reaction (25) also occurs.

$$[CH_3\overset{\displaystyle O}{\overset{\displaystyle \|}{C}}-Mn(CO)_5] \rightarrow [CH_3-Mn(CO)_5] + CO \qquad (25)$$

Catalytic cycles are commonly used to represent the chemistry occurring in catalysed reactions. Figure 7.4 illustrates the catalytic cycle involved in the hydroformylation reaction as catalysed by $[RhH(CO)L_2]$. The overall reaction is (26). This important industrial reaction, known as the Oxo process, is used

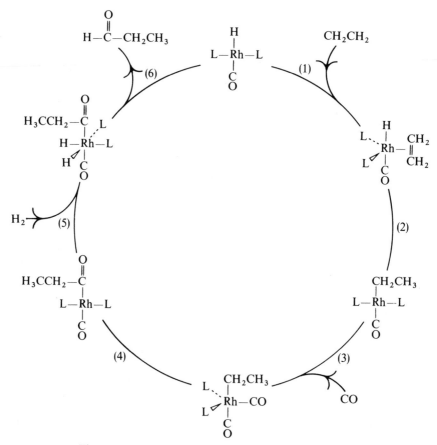

Figure 7.4   A proposed catalytic cycle for reaction (26).

$$CH_2{=}CH_2 + CO + H_2 \xrightarrow{\text{[RhH(CO)L}_2]} CH_3CH_2\overset{\displaystyle O}{\overset{\displaystyle \|}{C}}{-}H \qquad (26)$$

to produce a variety of aldehydes which can be reduced to alcohols; these are used to manufacture many useful products. In this cycle, the catalyst, a rhodium complex, undergoes a series of reactions and reappears unchanged at the end. The steps in the cycle illustrate the basic reactions which have just been introduced. In Figure 7.4: step (1) is an addition reaction; (2) is an insertion which is followed by (3), an addition to refill the coordination sites; (4) is another insertion; (5), an oxidative addition; and (6) a reductive elimination to produce product and regenerate the catalyst. The reactions are arranged in a cycle to emphasize the cyclic nature of this and of other catalytic processes. Such diagrams are common in the chemical and biochemical literature, since catalysis plays an important role in commercial processes and in life processes. Catalytic cycles are being developed to convert inexpensive simple molecules, such as CO and H_2, into a variety of more complicated and valuable compounds. We are also becoming increasingly aware of chemical cycles in living systems and the roles metals play in them.

7.3 Metals in living systems

A number of metals are essential to life. Extensive research has been done to establish the roles of these metals and the reactions and mechanisms through which they function. This field which studies metals in biology is called *bioinorganic chemistry*. Much of this work has been done recently and is a most exciting chemical frontier. Space will only allow a glance at this chemistry, and so we will focus on two classes of iron compounds whose nature has been revealed in recent years.

It has long been known that a number of metals (for example, magnesium and iron) are bound to the ligand, structure VI, in physiologically-important molecules. The ligand itself is known as porphine, and the metal complexes are called metal porphyrins. Only the basic ring structure is shown in structure VI, since it is found in a variety of modifications in living systems. The ligand has a charge of -2 and provides four nitrogen donor atoms which fit around a metal in a rigid square planar fashion. A metal bound in this way is also able to bind two other ligands, one above and one below the square plane.

One of the most familiar and intensively studied biochemical molecules is hemoglobin, a very large molecule (with a molecular weight of about 64,500), which contains four iron atoms each surrounded by a porphine ligand. This molecule readily binds and carries molecular oxygen from the lungs (where the partial pressure of O_2 is high) through the blood vessels to the sites in the body where oxygen is used in metabolic processes (and where the partial pressure of

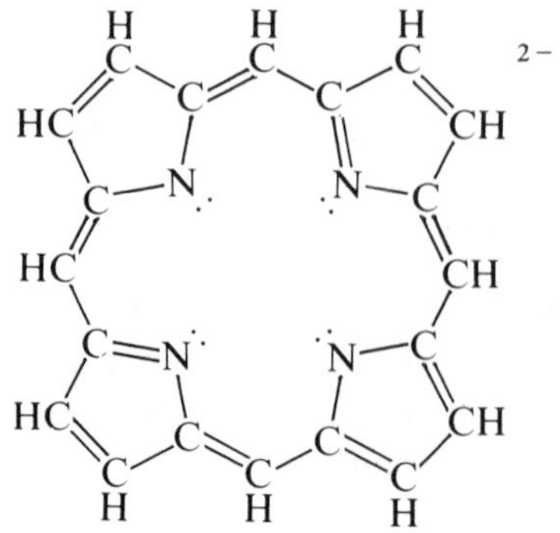

Porphine
anion

VI

O_2 is low). Hemoglobin contains iron(II). The complex without O_2 is five-coordinate and high-spin with four unpaired electrons. When O_2 is bound to it, the iron becomes low spin and the oxyhemoglobin complex is diamagnetic. In both cases, there is a nitrogen atom from an imidazole of the globin filling a fifth coordination site on the iron, and thus the O_2 containing complex is octahedral. It is rather surprising that the iron(II) is not oxidized by the O_2 which is a good oxidizing agent.

Because of the overall complexity of hemoglobin, chemists have long tried to make model compounds which would mimic the oxygen-carrying properties of the natural protein but which would be easier to prepare and study. The simplest model compounds contain iron(II) bound in a porphyrin ring with a nitrogen donating molecule in the fifth coordination site, structure VII.

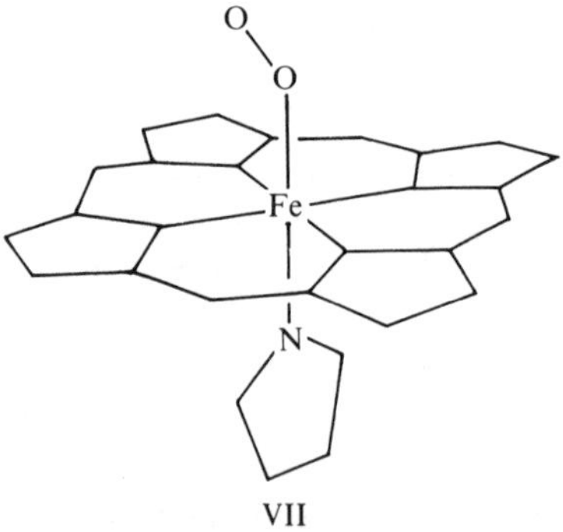

VII

Such compounds readily bind O_2; however, unless they are kept at very low temperatures the iron is rapidly oxidized to iron(III). The final products of these irreversible oxidations are generally the very stable μ-oxo bridged iron(III) species, reaction (27).

$$2[\text{LFe(por)}] + 1/2O_2 \xrightarrow{} [\text{L(por)Fe–O–Fe(por)L}] \qquad (27)$$

(L is an uncharged ligand and por is a −2 charged porphine group.) It was, therefore, necessary to design an iron complex which cannot form a stable μ-oxo bridged species. This was elegantly achieved with the synthesis of the "picket-fenced", structure VIII, and "capped" porphyrin, structure IX, iron(II)

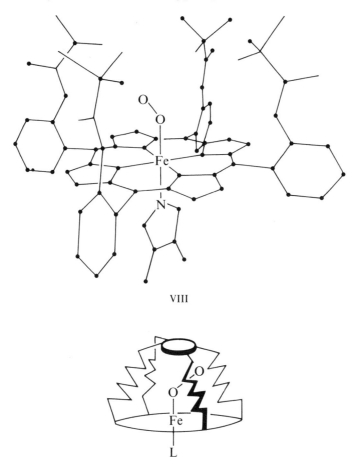

VIII

IX

complexes. The presence of large organic groups bonded to the top of the porphyrin ring allows small molecules such as O_2, CO, and NO to bind, but not large ligands such as imidazole or pyridine. If the axial coordination site, *trans* to the picket fence or the cap, is protected with a ligand, then the complex does not form a μ-oxo bridge and it efficiently acts as an oxygen carrier.

Another class of molecules which contain iron porphyrins is the cytochromes. These molecules serve as oxidizing agents when the iron is iron(III), and as reducing agents when it is iron(II), and are used to transport the energy derived from metabolism. In cytochrome *c*, the coordination sites above and below the iron are occupied by donor atoms (an N and an S) from other parts of the molecule, and the octahedrally-coordinated iron is low spin in both the $+2$ and $+3$ states. The facts that: (1) no spin change occurs during redox reactions; and (2) the rigid porphyrin structure makes the surroundings for iron(II) and iron(III) very similar explain their rapid electron transfer reactions. (Recall the Franck–Condon principle in Section 6.8.)

Other ligand frameworks around an iron ion are also suitable to produce rapid effective electron and energy transfer systems. One of the most interesting of these is in the ferredoxins. A particular ferredoxin has a molecular weight of about 6,000 and contains two independent Fe_4 units in which the irons are linked into a cubic cage by sulfur atoms and are also bonded to another sulfur in an aminoacid, structure X.

X

In this case, each iron has a distorted tetrahedral environment and both the iron(II) and (III) forms are high spin. It has been possible to characterize well the structure of the metal portion of these molecules, because it was found that the iron–sulfur cluster could be extruded from the natural molecule, the cluster could be independently synthesized, and the synthetic cluster inserted back to regenerate the original molecule and its biological activity.

A further development of this research was the discovery that the active site in nitrogen fixing bacteria is also a related iron–sulfur cluster. The significant differences are that one iron is replaced by a molybdenum and that the molybdenums from two of these clusters are linked by three bridging sulfur atoms. It is not uncommon that natural systems are able to produce quite varied chemistry by using some basic molecules which differ from one another in relatively minor ways.

7.4   Solid-state chemistry

The importance of solid-state electrical devices has led to a resurgence of interest in solid-state physics and chemistry. One of the best examples of interesting coordination chemistry in the solid state is in the field of one-dimensional electrical conductors. Square planar complexes crystallize in a variety of ways. One mode of crystal packing creates a stack of the square planar ions or molecules, one on top of the other. Magnus's green salt, $[Pt(NH_3)_4][PtCl_4]$, has been known as an anomaly for more than a century. Its grass green color differs markedly from its colorless cation and its red anion. This is now known to result from the one-dimensional interaction between the stacks of square planar counter ions.

The colorless complex $K_2[Pt(CN)_4]$ crystallizes in this way to form a nonconducting solid, structure XI.

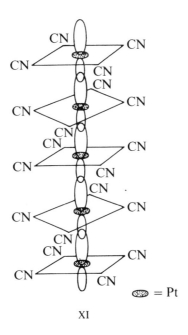

XI

In square planar complexes, the d_{z^2} orbital of the metal is viewed as lying perpendicular to the square plane, and in these complexes the d_{z^2} orbital on one metal is directed at the d_{z^2} orbitals on the metals above and below it. Since in this, and most other square planar complexes, the d_{z^2} orbitals are filled, repulsion occurs. The platinum–platinum separation in this complex is 348 pm (3.48 Å) in comparison with 278 pm (2.78 Å) in platinum metal. However, if the complex is oxidized, electrons are removed from these d_{z^2} orbitals, and a bonding interaction results. Partial oxidation of $K_2[Pt(CN)_4]$ results in the Pt atoms moving closer together (the separation becomes 288 pm (2.88 Å), and the crystals taking on intense colors and become electrical conductors. An interesting feature of the conductivity is that it occurs only in the direction perpendicular to the square planes of the $[Pt(CN)_4]^{2-}$ ions. Along that direction the interacting d_{z^2} orbitals, one on top of the other, form a delocalized molecular orbital from one side of the crystal to the other. Electron flow can occur through this partly filled orbital. The resulting substance is an example of a nonstoichiometric compound and it has the formula

$K_2[Pt(CN)_4Br_{0.3}] \cdot 3H_2O$, when the oxidizing agent is bromine. This complex, known as Klogman's salt, is the classical example of a low-dimensional solid conductor.

Another one-dimensional conductor is [Ni(tetramethylporphine)(I)]. It is made by oxidation of the square planar d^8 complex [Ni(tetramethylporphine)] with iodine. Again, this creates vacancies in overlapping d_{z^2} orbitals. The ultimate goal of such research is to discover solids which will be superconducting at 25°C. Solid-state inorganic chemistry has come of age and is an important area of research.

Problems

1. Write appropriate balanced equations and give approximate experimental conditions for the preparation of each of the following:

$Re(CO)_5Cl$ starting with $Re_2(CO)_{10}$
$Fe(C_5H_5)_2$ starting with $FeCl_2$
$K[PtCl_3(C_2H_4)]$ starting with $K_2[PtCl_4]$

2. Complete and balance the following equations.

$[PtCl_2(PCl_3)_2] + CH_3OH \rightarrow (CH_3OH$ solvent)
$Fe(CO)_5 + O_2 \rightarrow$ (gaseous reaction)

$Cr(C_6H_6)_2 \xrightarrow{\text{heat}}$ (no solvent)

3. Classify reactions (13) and (14) in Chapter 7 as (*a*) addition, (*b*) oxidative–addition, (*c*) reductive–elimination, or (*d*) insertion reactions.

4. $[Rh(OH)(CO)L_2] \xrightarrow[(1)]{+CO} [Rh(OH)(CO_2)L_2]$

$(6) \uparrow -H_2$ $(2) \downarrow$

$[Rh(H)_2(OH)(CO)L_2]$ $[Rh(CO_2H)(CO)L_2]$

$(5) \uparrow +H_2O$ $(3) \downarrow$

$[RhH(CO)L_2] \xleftarrow[(4)]{-CO_2} [RhH(CO_2)(CO)L_2]$

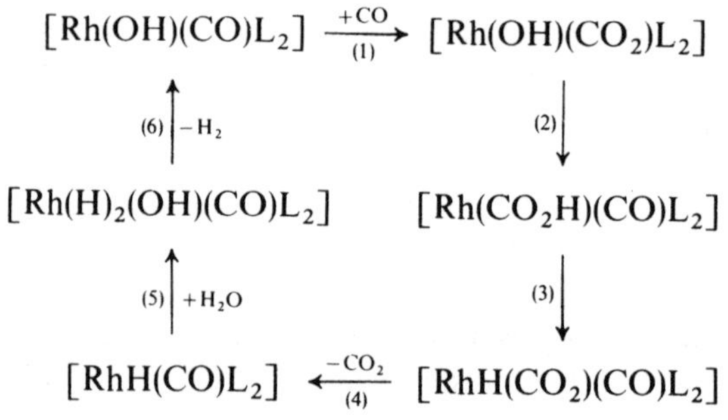

The series of reactions above is a simple catalytic cycle for the water gas shift reaction, $CO + H_2O \rightarrow CO_2 + H_2$. (*a*) What is the oxidation state of rhodium in each complex in the cycle? (*b*) Indicate whether each complex is a sixteen- or eighteen-electron complex. (*c*) Which of these classes of reactions does each step of the cycle illustrate: addition, elimination, oxidative–addition, reductive–elimination, insertion, or deinsertion?

5. Write a plausible catalytic cycle for the hydrogenation of ethene, $CH_2CH_2 + H_2 \rightarrow CH_3CH_3$, as catalysed by $[RhClL_3]$. Use only sixteen- and eighteen-electron complexes in the cycle, and only the types of reaction listed in problem 3(c). Note that $[RhClL_3]$ is called Wilkinson's catalyst, and that it is often used as a homogeneous hydrogenation catalyst.

6. Consider reaction (27) in Chapter 7. What are the oxidation states of iron in $[LFe(por)]$ and in $[L(por)Fe\text{–}O\text{–}Fe(por)L]$? If the mechanism for reaction (27) involves a collision between $[LFe(por)O_2]$ and $[LFe(por)]$, explain how a "picket fence" on the porphine ligand could prevent the reaction.

7. If the tetramethylporphine (tmp) ligand has a charge of -2, what is the oxidation state of nickel in $[Ni(tmp)]$ and $[Ni(tmp)(I_3)_{0.33}]$? How many electrons would be in the nickel $3d_{z^2}$ orbital in each complex?

References

Collman, J. P. and Hegedus, L. S. *Principles and Applications of Organotransition Metal Chemistry* (University Science Books, Mill Valley, California, 1980).

Eisch, J. J. and King, R. B. *Organometallic Syntheses* (Academic, New York, 1981).

Kochi, J. K. *Organometallic Mechanisms and Catalysis* (Academic, New York, 1978).

Parshall, G. W. *Homogeneous Catalysis, The Applications and Chemistry of Catalysis by Soluble Transitions Metal Complexes* (Wiley, New York, 1980).

Hughes, M. N. *The Inorganic Chemistry of Biological Processes*, 2nd edn (Wiley, New York, 1981).

Eichhorn, G. L. (ed.) *Inorganic Biochemistry* (Elsevier, Amsterdam, 1973).

Eichhorn, G. L. and Marzilli, L. (eds) *Advances in Inorganic Biochemistry*, 5 vols (Elsevier, Amsterdam, 1979–1983).

Ibers, J. A., Pace, L. J., Martinsen, J. and Hoffman, B. M. "Stacked Metal Complexes: Structures and Properties", *Structure and Bonding*, **50**, 3 (1982).

Index of Complexes

138

Index